Ulrich W. Kulisch

Advanced Arithmetic
for the Digital Computer

Design of Arithmetic Units

SpringerWienNewYork

Dr. Ulrich W. Kulisch
Professor of Mathematics
Institut für Angewandte Mathematik, Universität Karlsruhe, Germany

QA
76.9
.C62
K85
2002

This work is subject to copyright. All rights are reserved, whether the whole or part of the material is concerned, specifically those of translation, reprinting, re-use of illustrations, broadcasting, reproduction by photocopying machines or similar means, and storage in data banks.

© 2002 Springer-Verlag Wien
Printed in Austria

Product Liability: The publisher can give no guarantee for all the information contained in this book. This does also refer to information about drug dosage and application thereof. In every individual case the respective user must check its accuracy by consulting other pharmaceutical literature. The use of registered names, trademarks, etc. in this publication does not imply, even in the absence of a specific statement, that such names are exempt from the relevant protective laws and regulations and therefore free for general use.

Cover illustration: Falko Schröder, University of Applied Science, Zweibrücken
Typesetting: Camera ready by editor
Printing and binding: MANZ CROSSMEDIA, Vienna
Printed on acid-free and chlorine-free bleached paper
SPIN: 10892255
CIP data applied for

With 32 Figures (1 in colour)

ISBN 3-211-83870-8 Springer-Verlag Wien New York

Preface

The number one requirement for computer arithmetic has always been speed. It is the main force that drives the technology. With increased speed larger problems can be attempted. To gain speed, advanced processors and programming languages offer, for instance, compound arithmetic operations like *matmul* and *dotproduct*.

But there is another side to the computational coin - the accuracy and reliability of the computed result. Progress on this side is very important, if not essential. Compound arithmetic operations, for instance, should always deliver a correct result. The user should not be obliged to perform an error analysis every time a compound arithmetic operation, implemented by the hardware manufacturer or in the programming language, is employed.

This treatise deals with computer arithmetic in a more general sense than usual. *Advanced computer arithmetic* extends the accuracy of the elementary floating-point operations, for instance, as defined by the IEEE arithmetic standard, to all operations in the usual product spaces of computation: the complex numbers, the real and complex intervals, and the real and complex vectors and matrices and their interval counterparts. The implementation of advanced computer arithmetic by fast hardware is examined in this book. Arithmetic units for its elementary components are described. It is shown that the requirements for speed and for reliability do not conflict with each other. Advanced computer arithmetic is superior to other arithmetic with respect to accuracy, costs, and speed.

Vector processing is an important technique used to speed up computation. Difficulties concerning the accuracy of conventional vector processors are addressed in [116, 117]. See also [32] and [78]. Accurate vector processing is subsumed in what is called advanced computer arithmetic in this treatise. Compared with elementary floating-point arithmetic it speeds up computations considerably and it eliminates many rounding errors and exceptions. Its implementation requires little more hardware than is needed for elementary floating-point arithmetic. All this strongly supports the case for implementing such advanced computer arithmetic on every CPU. With the speed computers have reached and the problem sizes that are dealt with, vector operations should be performed with the same reliability as elementary floating-point operations.

On parallel computers faster and more powerful arithmetic units essentially reduce the number of processors needed and the complexity of the interconnection network. Thus a desired efficiency can be reached at a lower cost.

A basic feature of advanced computer arithmetic as well as of vector processing is the two instructions *accumulate* and *multiply and accumulate* added to the instruction set for floating-point numbers. The first instruction is a particular case of the second one, which computes a sum of products - the dot product or scalar product of two vectors. Pipelining makes these operations really fast. We show in the first chapter that fixed-point accumulation of products is the fastest way to accumulate scalar products on a computer. This is so for all kinds of computers - personal computers, workstations, mainframes or super computers. In contrast to floating-point accumulation of products, fixed-point accumulation is error free. Not a single bit is lost. The new operation is gained at modest cost. It increases both the speed of a computation as well as the accuracy of the computed result.

A conventional floating-point computation may fail to produce a correct answer without any error signal being given to the user. A very worthy goal of computing, therefore, would be to do rigorous mathematics with the computer. Examples of such rigour are verified solution of differential or integral equations, or validation of the solution of a system of equations with proof of existence and uniqueness of the solution within the computed bounds. Interval arithmetic serves this purpose. If the verification or validation step fails, the user is made aware that some more powerful tool has to be applied. Higher precision arithmetic might then be used, for instance. Variable precision arithmetic is thus a sound complement for interval arithmetic. With a fast and accurate scalar product, fast multiple precision arithmetic can be easily provided on the computer.

The second chapter of this booklet reveals a necessary and sufficient condition under which a computer representable element in any one of the relevant computational spaces has a unique additive inverse.

The third chapter deals with interval arithmetic. It is shown that on superscalar processors the four basic interval operations can be made as fast as simple floating-point operations with only modest additional hardware costs. In combination with the results of the first chapter - a hardware-supported accurate scalar product - interval vector and matrix operations can be performed with highest accuracy and faster than with simple floating-point arithmetic.

The three chapters of this volume were written as independent articles. They were prepared while the author was staying at the Electrotechnical Laboratory (ETL), Agency of Industrial Science and Technology, MITI, at Tsukuba, Japan, during sabbaticals in 1998 and in 1999/2000. Gathering these articles into a single publication raised the question of whether the text should be rewritten and reorganized into a unitary exposition. Thus a small number of repetitions - of simple definitions, of historic remarks, or of the list of references - could have been avoided. I decided not to do so. The articles have been prepared to help implement different aspects of advanced arithmetic on the computer. So it seems preferable not to interweave and combine separate things into a complex whole. Readability should have the highest priority.

I am grateful to all those colleagues and co-workers who have contributed through their research to the development of advanced computer arithmetic as it is presented in this treatise. In particular I would like to mention and thank Gerd Bohlender, Willard L Miranker, Reinhard Kirchner, Siegfried M.Rump, Thomas Teufel, Harald Böhm, Jürgen Wolff von Gudenberg, Andreas Knöfel, and Christof Baumhof.

I gratefully acknowledge the help of Neville Holmes who went carefully through the manuscripts, sending back corrections and suggestions that led to many improvements.

Finally I wish to thank the Electrotechnical Laboratory, Agency of Industrial Science and Technology at Tsukuba, Japan for providing me the opportunity to write the articles in a pleasant scientific environment without constantly being interrupted by the usual university business. I especially owe thanks to Satoshi Sekiguchi for being a wonderful host personally and scientifically. I am looking forward to, and eagerly await, advanced computer arithmetic on commercial computers.

Karlsruhe, July 2002   *Ulrich W. Kulisch*

The picture on the cover page illustrates the contents of the book. It is showing a chip for fast Advanced Computer Arithmetic and eXtended Precision Arithmetic (ACA-XPA). Its components are symbolically indicated on top: hardware support for 15 basic arithmetic operations including accurate scalar products with different roundings and case selections for interval multiplication and division. Corresponding circuits are developed in the book.

The picture is showing friends with Ursula Kulisch flanked by the host, Satoshi Sekiguchi, and Ulrich Kulisch.

Contents

1. **Fast and Accurate Vector Operations** 1
   1.1 Introduction 1
       1.1.1 Background 1
       1.1.2 Historic Remarks................................. 7
   1.2 Implementation Principles 10
       1.2.1 Solution A: Long Adder and Long Shift 13
       1.2.2 Solution B: Short Adder with Local Memory on the
             Arithmetic Unit 14
       1.2.3 Remarks 15
       1.2.4 Fast Carry Resolution............................. 17
   1.3 High-Performance Scalar Product Units (SPU) 19
       1.3.1 SPU for Computers with a 32 Bit Data Bus 19
       1.3.2 SPU for Computers with a 64 Bit Data Bus 23
   1.4 Comments on the Scalar Product Units 25
       1.4.1 Rounding 25
       1.4.2 How much Local Memory should be Provided on a SPU? 27
       1.4.3 A SPU Instruction Set 28
       1.4.4 Interaction with High Level Programming Languages . 30
   1.5 Scalar Product Units for Top-Performance Computers....... 32
       1.5.1 Long Adder for 64 Bit Data Word (Solution A) 32
       1.5.2 Long Adder for 32 Bit Data Word (Solution A) 37
       1.5.3 Short Adder with Local Memory on the Arithmetic
             Unit for 64 Bit Data Word (Solution B) 40
       1.5.4 Short Adder with Local Memory on the Arithmetic
             Unit for 32 Bit Data Word (Solution B) 45
   1.6 Hardware Accumulation Window 49
   1.7 Theoretical Foundation of Advanced Computer Arithmetic .. 53
   Bibliography and Related Literature.......................... 63

2. **Rounding Near Zero** 71
   2.1 The one dimensional case................................ 71
   2.2 Rounding in product spaces.............................. 75
   Bibliography and Related Literature.......................... 79

3. Interval Arithmetic Revisited 81
3.1 Introduction and Historical Remarks 82
3.2 Interval Arithmetic, a Powerful Calculus to Deal with Inequalities 89
3.3 Interval Arithmetic as Executable Set Operations 92
3.4 Enclosing the Range of Function Values 97
3.5 The Interval Newton Method 101
3.6 Extended Interval Arithmetic 104
3.7 The Extended Interval Newton Method 110
3.8 Differentiation Arithmetic, Enclosures of Derivatives 112
3.9 Interval Arithmetic on the Computer 116
3.10 Hardware Support for Interval Arithmetic 127
    3.10.1 Addition $A+B$ and Subtraction $A-B$ 128
    3.10.2 Multiplication $A*B$ 128
    3.10.3 Interval Scalar Product Computation 131
    3.10.4 Division A/B 133
    3.10.5 Instruction Set for Interval Arithmetic 134
    3.10.6 Final Remarks 135
Bibliography and Related Literature 137

1. Fast and Accurate Vector Operations

Summary.
Advances in computer technology are now so profound that the arithmetic capability and repertoire of computers can and should be expanded. Nowadays the elementary floating-point operations $+, -, \times, /$ give computed results that coincide with the rounded exact result for any operands. Advanced computer arithmetic extends this accuracy requirement to all operations in the usual product spaces of computation: the real and complex vector spaces as well as their interval correspondents. This enhances the mathematical power of the digital computer considerably. A new computer operation, the scalar product, is fundamental to the development of advanced computer arithmetic.

This paper studies the design of arithmetic units for advanced computer arithmetic. Scalar product units are developed for different kinds of computers like personal computers, workstations, mainframes, super computers or digital signal processors. The new expanded computational capability is gained at modest cost. The units put a methodology into modern computer hardware which was available on old calculators before the electronic computer entered the scene. In general the new arithmetic units increase both the speed of computation as well as the accuracy of the computed result. The circuits developed in this paper show that there is no way to compute an approximation of a scalar product faster than the correct result.

A collection of constructs in terms of which a source language may accommodate advanced computer arithmetic is described in the paper. The development of programming languages in the context of advanced computer arithmetic is reviewed. The simulation of the accurate scalar product on existing, conventional processors is discussed. Finally the theoretical foundation of advanced computer arithmetic is reviewed and a comparison with other approaches to achieving higher accuracy in computation is given. Shortcomings of existing processors and standards are discussed.

1.1 Introduction

1.1.1 Background

Advances in computer technology are now so profound that the arithmetic capability and repertoire of computers can and should be expanded. At a time when more than 100 million transistors can be placed on a single chip, computing speed is measured in giga- and teraflops, and memory space in giga-words, there is no longer any need to perform all computer calculations by the four elementary floating-point operations with all the shortcomings of this arithmetic (for the shortcomings see the three examples listed in Section 1.1.2).

Nowadays the elementary floating-point operations $+, -, \times, /$ give computed results that coincide with the rounded exact result of the operation for any operands. See, for instance, the IEEE-Arithmetic Standards 754 and 854, [114, 115]. Advanced computer arithmetic extends this accuracy requirement to all operations in the usual product spaces of computation: the complex numbers, real and complex vectors, real and complex matrices, real and complex intervals as well as real and complex interval vectors and interval matrices. This enhances the mathematical power of the digital computer considerably. A great many computer operations can then be performed with but a single rounding error.

If, for instance, the scalar product of two vectors with 1000 components is to be computed about 2000 roundings are executed in conventional floating-point arithmetic. Advanced arithmetic reduces this to a single rounding. The computed result is within a single rounding error of the correct result.

The new operations are distinctly different from the customary ones which are based on elementary floating-point arithmetic. A careful analysis and a general theory of computer arithmetic [60, 62] show that the new operations can be built up on the computer by a modular technique as soon as a new fundamental operation, the scalar product, is provided with full accuracy on a low level, possibly in hardware.

The computer realization of the scalar product of two floating-point vectors can be achieved with full accuracy in several ways. A most natural way is to add the products of corresponding vector components into a long fixed-point register (accumulator) which covers twice the exponent range of the floating-point format in which the vector components are given. Use of the long accumulator has the advantage of being rather simple, straightforward and fast. Since fixed-point accumulation of numbers is error free it always provides the desired accurate answer. The technique was already used on old mechanical calculators long before the electronic computer.

In a floating-point system the number of mantissa digits and the exponent range are finite. Therefore, the fixed-point register is finite as well, and it is relatively small, consisting of about one to four thousand bits depending on the data format in use. So we have the seemingly paradoxing and striking situation that scalar products of floating-point vectors with even millions of components can be computed to a fully accurate result using a relatively small finite local register on the arithmetic unit.

In numerical analysis the scalar or dot product is ubiquitous. It is not merely a fundamental operation in all the product spaces mentioned above.

The process of residual or defect correction, or of iterative refinement, is composed of scalar products. There are well known limitations to these processes in floating-point arithmetic. The question of how many digits of a defect can be guaranteed with single, double or extended precision arithmetic has been carefully investigated. With the optimal scalar product the defect can always be computed to full accuracy. It is the accurate scalar product which makes residual correction effective.

With the accurate scalar product quadruple or multiple precision arithmetic can easily be provided on the computer. This enables the user to use higher precision operations in numerically critical parts of his computation. It helps to increase software reliability. A multiple precision number is represented as an array of floating-point numbers. The value of this number is the sum of its components. It can be represented in the long accumulator. Addition and subtraction of multiple precision variables or numbers can easily be performed in the long accumulator. Multiplication of two such numbers is simply a sum of products. It can be computed by means of the accurate scalar product. For instance in case of a fourfold precision the product of two such numbers $a = (a_1 + a_2 + a_3 + a_4)$ and $b = (b_1 + b_2 + b_3 + b_4)$ is obtained by

$$\begin{aligned} a \times b &= (a_1 + a_2 + a_3 + a_4) \times (b_1 + b_2 + b_3 + b_4) \\ &= a_1 b_1 + a_1 b_2 + a_1 b_3 + a_1 b_4 + a_2 b_1 + \cdots + a_4 b_3 + a_4 b_4 \\ &= \sum_{i=1}^{4} \sum_{j=1}^{4} a_i b_j. \end{aligned}$$

Using the long accumulator the result is independent of the sequence in which the summands are added. For details see Remark 3 on page 60 in section 1.7.

Approximation of a continuous function by a polynomial by the method of least squares leads to the Hilbert matrix as coefficients. It is extremely ill conditioned. It is well known that it is impossible to invert the Hilbert matrix in double precision floating-point arithmetic successfully by any direct or iterative method for dimensions greater than 11. Implementation of the accurate scalar product in hardware also supports very fast multiple precision arithmetic. It easily inverts the Hilbert matrix of dimension 40 to full accuracy on a PC in a very short computing time. If increase or decrease of the precision in a program is provided by the programming environment, the user or the computer itself can choose the precision which optimally fits his problem.

Inversion of the Hilbert matrix of dimension 40 is impossible with quadruple precision arithmetic. With it only one fixed precision is available. If one runs out of precision in a certain problem class, one often runs out of quadruple precision very soon as well. It is preferable and simpler, therefore, to provide the principles for enlarging the precision than simply providing any fixed higher precision. A hardware implementation of a full quadruple precision arithmetic is much more costly than an implementation of the accurate scalar product. The latter only requires fixed-point accumulation of the products. On the computer, there is only one standardized floating-point format that is double precision.

With increasing speed of computers, problems to be dealt with become larger. Instead of two dimensional problems users would like to solve three dimensional problems. Gauss elimination for a linear system of equations requires the magnitude of $\mathcal{O}(n^3)$ operations. Large, sparse or structured linear or non linear systems, therefore, can only be solved iteratively. The basic operation of iterative methods (Jacobi method, Gauss-Seidel method, overrelaxation method, conjugate gradient method, Krylov space methods, multigrid methods and others like the QR method for the computation of eigenvalues) is the matrix-vector multiplication which consists of a number of scalar products. It is well known that finite precision arithmetic often worsens the convergence of these methods. An iterative method which converges to the solution in infinite precision arithmetic often converges much slower or even diverges in finite precision arithmetic. The accurate scalar product is faster than a computation in conventional floating-point arithmetic. In addition to that it can speed up the rate of convergence of iterative methods significantly in many cases [27, 28].

For many applications it is necessary to compute the value of the derivative of a function. Newton's method in one or several variables is a typical example for this. Modern numerical analysis solves this problem by automatic or algorithmic differentiation. The so called reverse mode is a very fast method of automatic differentiation. It computes the gradient, for instance, with at most five times the number of operations which are needed to compute the function value. The memory overhead and the spatial complexity of the reverse mode can be significantly reduced by the exact scalar product if this is considered as a single, always correct, basic arithmetic operation in the vector spaces [88]. The very powerful methods of global optimization [79], [80], [81] are impressive applications of these techniques.

Many other applications require that rigorous mathematics can be done with the computer using floating-point arithmetic. As an example, this is essential in simulation runs (fusion reactor, eigenfrequencies of large generators) or mathematical modelling where the user has to distinguish between computational artifacts and genuine reactions of the model. The model can only be developed systematically if errors resulting from the computation can be excluded.

Nowadays computer applications are of immense variety. Any discussion of where a dot product computed in quadruple or extended precision arithmetic can be used to substitute for the accurate scalar product is superfluous. Since the former can fail to produce a correct answer an error analysis is needed for all applications. This can be left to the computer. As the scalar product can always be executed correctly with moderate technical effort it should indeed always be executed correctly. An error analysis thus becomes irrelevant. Furthermore, the same result is always obtained on different computer platforms. A fully accurate scalar product eliminates many rounding errors in numerical computations. It stabilizes these computations and speeds them up as well. It is the necessary complement to floating-point arithmetic.

This paper studies the design of arithmetic units for advanced computer arithmetic. Scalar product units are developed for different kinds of computers like personal computers, workstations, mainframes, super computers or even digital signal processors. The differences in the circuits for these diverse processors are dictated by the speed with which the processor delivers the data to the arithmetic or scalar product unit. The data are the vector components. In all cases the new expanded computational capability is gained at modest cost. The cost increase is comparable to that from a simple to a fast multiplier, for instance, by a Wallace tree, accepted years ago. It is a main result of our study that for all processors mentioned above circuits can be given for the computation of the accurate scalar product with virtually no computing time needed for the execution of the arithmetic. In a pipeline, the arithmetic can be executed within the time the processor needs to read the data into the arithmetic unit. This means, that no other method to compute a scalar product can be faster, in particular not a conventional approximate computation of the scalar product in floating-point arithmetic which can lead to an incorrect result.

In the pipeline a multiplication and the accumulation of a product to the intermediate sum in the long accumulator are performed simultaneously. This doubles the speed of the accurate scalar product in comparison with a conventional computation in floating-point arithmetic where these operations are performed sequentially. Furthermore, fixed-point accumulation of the products is simpler than accumulation in floating-point. Many intermediate steps that are executed in a floating-point accumulation such as normalization and rounding of the products and the intermediate sum, composition into a floating-point number and decomposition into mantissa and exponent for the next operation do not occur in the fixed-point accumulation of the accurate scalar product used in advanced computer arithmetic.

In recent years there has been a significant shift of numerical computation from general purpose computers towards vector and parallel computers – so-called super computers. Along with the four elementary floating-point operations these computers usually offer compound operations as additional arithmetic operations. A particular such compound operation, *multiply and accumulate*, is provided for the computation of the scalar product of two vectors. These compound operations are heavily pipelined and make the computation really fast. They are automatically inserted in a user's program by a vectorizing compiler. However, if these operations are not carefully implemented the user loses complete control of his computation.

In 1987 GAMM[1] and IMACS[2] published a *Resolution on Computer Arithmetic* [116] which criticized the mathematically inadequate execution of matrix and vector operations on all existing vector processors. An amendment was demanded. The user should not be obliged to perform an error analysis every time an elementary compound operation, predefined by the manufacturer, is employed. In 1993 the two organizations approved and published a *Proposal for Accurate Floating-Point Vector Arithmetic* [117]. It requires a mathematically correct implementation of matrix and vector operations, in particular, of the accurate scalar product on **all** computers. In 1995 the IFIP-Working Group 2.5 on Numerical Software endorsed this proposal. Meanwhile it became an EU Guideline.

We finish this Section with a warning to the reader. This chapter does not consist of independent sections. The later sections are built upon the earlier ones. On the other hand material that is presented later can be helpful in contributing to a full understanding of circuits that are discussed earlier. In Section 1.2 basic ideas for the solution of the problem are discussed. Section 1.3 develops fast solutions for small and medium size computers. Section 1.5 then considers solutions for very fast systems.

New fundamental computer operations must be embedded into programming languages where they can be activated by the user. Of course, operator notation is the ultimate solution. However, it turns out to be extremely useful to put a number of elementary instructions into the hands of the user also. Such instructions for low and high level languages are discussed in Section 1.4, in particular in 1.4.3 and 1.4.4. These are based on a long lasting experience with the XSC-languages since 1980. Some readers might not so much be interested in the very fast solutions in Section 1.5. Therefore, the interplay with programming languages is already presented in Section 1.4.

This text summarizes both an extensive research activity during the past twenty years and the experience gained through various implementations of the entire arithmetic package on diverse processors. The text is also based on lectures held at the Universität Karlsruhe during the preceding 25 years. While the collection of research articles that contribute to this paper is not very large in number, I refrain from a detailed review of them and refer the reader to the list of references. This text synthesizes and organizes diverse contributions into a coherent presentation. In many cases more detailed information can be obtained from original doctoral theses.

---

[1] GAMM = Gesellschaft für Angewandte Mathematik und Mechanik
[2] IMACS = International Association for Mathematics and Computers in Simulation

1.1.2 Historic Remarks

Floating-point arithmetic has been used since the early forties and fifties (Zuse Z3, 1941) [11, 82]. Technology in those days was poor (electromechanical relays, electron tubes). It was complex and expensive. The word size of the Z3 consisted of 24 bits. The storage provided 64 words. The four elementary floating-point operations were all that could be provided. For more complicated calculations an error analysis was left to and put on the shoulder of the user.

Before that time, highly sophisticated mechanical computing devices were used. Several very interesting techniques provided the four elementary operations addition, subtraction, multiplication and division. Many of these calculators were able to perform an additional *fifth operation* which was called *Auflaufenlassen* or the *running total*. The input register of such a machine had perhaps 10 or 12 decimal digits. The result register was much wider and had perhaps 30 digits. It was a fixed-point register which could be shifted back and forth relative to the input register. This allowed a continuous accumulation of numbers and of products of numbers into different positions of the result register. Fixed-point accumulation is thus error free. See Fig. 1.22 and Fig. 1.23 on page 62. *This fifth arithmetic operation was the fastest way to use the computer. It was applied as often as possible. No intermediate results needed to be written down and typed in again for the next operation. No intermediate roundings or normalizations had to be performed. No error analysis was necessary. As long as no under- or overflow occurred, which would be obvious and visible, the result was always correct. It was independent of the order in which the summands were added. If desired, only one final rounding was executed at the very end of the accumulation.*

This extremely useful and fast fifth arithmetic operation was not built into the early floating-point computers. It was too expensive for the technologies of those days. Later its superior properties had been forgotten. Thus floating-point arithmetic is still somehow incomplete.

After Zuse, the early electronic computers in the late forties and early fifties represented their data as fixed-point numbers. Fixed-point arithmetic was used because of its superior properties. Fixed-point addition and subtraction are error free. Fixed-point arithmetic with a rather limited word size, however, imposed a scaling requirement. Problems had to be preprocessed by the user so that they could be accommodated by this fixed-point number representation. With increasing speed of computers, the problems that could be solved became larger and larger. The necessary preprocessing soon became an enormous burden.

Thus floating-point arithmetic became generally accepted. It largely eliminated this burden. A scaling factor is appended to each number in floating-point representation. The arithmetic itself takes care of the scaling. An exponent addition (subtraction) is executed during multiplication (division). It may result in a big change in the value of the exponent. But multiplication and division are relatively stable operations in floating-point arithmetic. Addition and subtraction, on the contrary, are troublesome in floating-point.

The quality of floating-point arithmetic has been improved over the years. The data format was extended to 64 and even more bits and the IEEE-arithmetic standard has finally taken the bugs out of particular realizations. Floating-point arithmetic has been used very successfully in the past. Very sophisticated and versatile algorithms and libraries have been developed for particular problems. However, in a general application the result of a floating-point computation is often hard to judge. It can be satisfactory, inaccurate or even completely wrong. The computation itself as well as the computed data do not indicate which one of the three cases has occurred. We illustrate the typical shortcomings by three very simple examples. All data in these examples are IEEE double precision floating-point numbers! For these and other examples see [84]:

1. Compute the following, theoretically equivalent expressions:

$$
\begin{array}{llllll}
10^{20} + & 17 - & 10 + & 130 - & 10^{20} \\
10^{20} - & 10 + & 130 - & 10^{20} + & 17 \\
10^{20} + & 17 - & 10^{20} - & 10 + & 130 \\
10^{20} - & 10 - & 10^{20} + & 130 + & 17 \\
10^{20} - & 10^{20} + & 17 - & 10 + & 130 \\
10^{20} + & 17 + & 130 - & 10^{20} - & 10
\end{array}
$$

A conventional computer using the data format double-precision of the IEEE floating-point arithmetic standard returns the values 0, 17, 120, 147, 137, -10. These errors come about because the floating-point arithmetic is unable to cope with the digit range required with this calculation. Notice that the data cover less than 4% of the digit range of the data format double precision!

2. Compute the solution of a system of two linear equations $Ax = b$, with

$$ A = \begin{pmatrix} 64919121 & -159018721 \\ 41869520.5 & -102558961 \end{pmatrix}, \quad b = \begin{pmatrix} 1 \\ 0 \end{pmatrix} $$

The solution can be expressed by the formulas:

$$ x_1 = a_{22}/(a_{11}a_{22} - a_{12}a_{21}) \quad \text{and} \quad x_2 = -a_{21}/(a_{11}a_{22} - a_{12}a_{21}) . $$

A workstation using IEEE double precision floating-point arithmetic returns the *approximate solution*:

$$ \tilde{x}_1 = 102558961 \quad \text{and} \quad \tilde{x}_2 = 41869520.5 , $$

while the correct solution is

$$x_1 = 205117922 \text{ and } x_2 = 83739041.$$

After only 4 floating-point operations all digits of the computed solution are wrong. A closer look into the problem reveals that the error happens during the computation of the denominator. This is just the kind of expression which always can be computed error free by the missing fifth operation.

3. Compute the scalar product of the two vectors a and b with five components each:

$$\begin{aligned}
a_1 &= 2.718281828 * 10^{10} & b_1 &= 1486.2497 * 10^9 \\
a_2 &= -3.141592654 * 10^{10} & b_2 &= 878366.9879 * 10^9 \\
a_3 &= 1.414213562 * 10^{10} & b_3 &= -22.37492 * 10^9 \\
a_4 &= 0.5772156649 * 10^{10} & b_4 &= 4773714.647 * 10^9 \\
a_5 &= 0.3010299957 * 10^{10} & b_5 &= 0.000185049 * 10^9
\end{aligned}$$

The correct value of the scalar product is $-1.00657107 * 10^8$. IEEE-double precision arithmetic delivers $+4.328386285 * 10^9$ so even the sign is incorrect. Note that no vector element has more than 10 decimal digits.

Problems that can be solved by computers become larger and larger. Today fast computers are able to execute several billion floating-point operations in each second. This number exceeds the imagination of any user. Traditional error analysis of numerical algorithms is based on estimates of the error of each individual arithmetic operation and on the propagation of these errors through a complicated algorithm. It is simply no longer possible to expect that the error of such computations can be controlled by the user. There remains no alternative to further develop the computer's arithmetic and to furnish it with the capability of control and validation of the computational process.

Computer technology is extremely powerful today. It allows solutions which even an experienced computer user may be totally unaware of. Floating-point arithmetic which may fail in simple calculations, as illustrated above, is no longer adequate to be used exclusively in computers of such gigantic speed for huge problems. The reintroduction of the fifth arithmetic operation, the accurate scalar product, into computers is a step which is long overdue. A central and fundamental operation of numerical analysis which can be executed correctly with only modest technical effort should indeed always be executed correctly and no longer only *approximately*. With the accurate scalar product all the nice properties which have been listed in connection with the old mechanical calculators return to the modern digital computer. *The accurate scalar product is the fastest way to use the computer. It should be applied as often as possible. No intermediate results need to be stored and read in again for the next operation. No intermediate roundings and normalizations have to be performed. No intermediate over- or underflow can occur.*

No error analysis is necessary. The result is always correct. It is independent of the order in which the summands are added. If desired, only one final rounding is executed at the very end of the accumulation.

This paper pleads for an extension of floating-point arithmetic by the accurate scalar product as fifth elementary operation. This combines the advantages of floating-point arithmetic (no scaling requirement) with those of fixed-point arithmetic (error free accumulation of numbers and of single products of numbers even for very long sums). It is obtained by putting a methodology into modern computer hardware which was already available on calculators before the electronic computer entered the scene.

To wipe out frequent misunderstandings a few words about what the paper does (and what it not does) seem to be necessary. The paper claims that, and explains how, scalar products, *the data of which are floating-point numbers*, can always be correctly computed. In the old days of computing (1950 – 1980) computers often provided sloppy arithmetic in order to be fast. This was "justified" by explaining that the last bit was incorrect in many cases, due to rounding errors. So why should the arithmetic be slowed down or more hardware be invested by computing the best possible answer of the operations under the assumption that the last bit is correct? Today it is often asked: why do we need an accurate scalar product? The last bit of the data is often incorrect and a floating-point computation of the scalar product delivers the best possible answer for problems with perturbed data. With the IEEE arithmetic standard this kind of "justification" has been overcome. This allows problems to be handled correctly where the data are correct and the best possible result of an operation is needed. With respect to the scalar product several such problems are listed in Section 1.1.1. In mathematics it makes a big difference whether a computation is correct for many or most data or for all data! For problems with perturbed (inexact) data interval arithmetic is the appropriate tool. Even in this case fixed-point accumulation of the scalar product delivers bounds or an approximation faster than a conventional computation in floating-point arithmetic. In summary, the paper extends the accuracy requirements of the IEEE arithmetic standard to the scalar product and with it to all operations in the usual product spaces of computation which are listed in the second paragraph of Section 1.1.1. All this is shortly called advanced computer arithmetic.

1.2 Implementation Principles

A normalized floating-point number x (in sign-magnitude representation) is a real number of the form $x = *m\,b^e$. Here $* \in \{+, -\}$ is the sign of the number, b is the base of the number system in use and e is the exponent. The base b is an integer greater than unity. The exponent e is an integer between two fixed integer bounds $e1$ and $e2$, and in general $e1 \leq 0 \leq e2$. The mantissa m is of the form

$$m = \sum_{i=1}^{l} d_i \, b^{-i}.$$

The d_i are the digits of the mantissa. They have the property $d_i \in \{0, 1, \ldots, b-1\}$ for all $i = 1(1)l$ and $d_1 \neq 0$. Without this last condition floating-point numbers are said to be unnormalized. The set of normalized floating-point numbers does not contain zero. For a unique representation of zero we assume the mantissa and the exponent to be zero. Thus a floating-point system depends on the four constants $b, l, e1$ and $e2$. We denote it by $R = R(b, l, e1, e2)$. Occasionally we shall use the abbreviations $\text{sign}(x)$, $\text{mant}(x)$ and $\exp(x)$ to denote the sign, mantissa and exponent of x respectively.

Nowadays the elementary floating-point operations $+, -, \times, /$ give computed results that coincide with the rounded exact result of the operation for any operands. See, for instance, the IEEE Arithmetic Standards 754 and 854, [114, 115]. Advanced computer arithmetic extends this accuracy requirement to all operations in the usual product spaces of computation: the complex numbers, the real and complex vectors, real and complex matrices, real and complex intervals as well as the real and complex interval vectors and interval matrices.

A careful analysis and a general theory of computer arithmetic [60, 62] show that all arithmetic operations in the computer representable subsets of these spaces can be realized on the computer by a modular technique as soon as fifteen fundamental operations are made available at a low level, possibly by fast hardware routines. These fifteen operations are

$$\boxplus, \boxminus, \boxtimes, \boxslash, \boxdot,$$
$$\triangledown, \triangledown, \triangledown, \triangledown, \triangledown,$$
$$\triangle, \triangle, \triangle, \triangle, \triangle.$$

Here $\square, * \in \{+, -, \times, /\}$ denotes (semimorphic[3]) operations using some particular monotone and antisymmetric rounding $\square: \mathbb{R} \to R$ such as rounding to the nearest floating-point number or rounding towards zero. Likewise $\triangledown$ and $\triangle, * \in \{+, -, \times, /\}$ denote the operations using the optimal (monotone[3]) rounding downwards $\triangledown: \mathbb{R} \to R$, and the optimal (monotone[3]) rounding upwards $\triangle: \mathbb{R} \to R$, respectively. $\boxdot, \triangledown$ and $\triangle$ denote scalar products with high accuracy. That is, if $a = (a_i)$ and $b = (b_i)$ are vectors with floating-point components, $a_i, b_i \in R$, then $a \odot b := \bigcirc (a_1 \times b_1 + a_2 \times b_2 + \ldots + a_n \times b_n)$, $\bigcirc \in \{\square, \triangledown, \triangle\}$. The multiplication and addition signs on the right hand side of the assignment denote exact multiplication and summation in the sense of real numbers.

---
[3] For a precise mathematical definition see Section 1.7.

These 15 operations are sufficient for the computer implementation of all arithmetic operations that are to be defined for all numerical data types listed above in the third paragraph of this section. Of the 15 fundamental operations, traditional numerical methods use only the four operations $\boxplus$, $\boxminus$, $\boxtimes$ and $\boxslash$. Interval arithmetic requires the eight operations $\triangledown$, $\triangledown$, $\triangledown$, $\triangledown$ and $\vartriangle$, $\vartriangle$, $\vartriangle$, $\vartriangle$. These eight operations are computer equivalents of the operations for real floating-point intervals, i. e. of interval arithmetic. Processors which support the IEEE arithmetic standard, for instance, offer 12 of these 15 operations: $\boxdot$, $\triangledown$, $\vartriangle$, $* \in \{+, -, \times, /\}$. The latter 8 operations $\triangledown$, $\vartriangle$, $* \in \{+, -, \times, /\}$ are not yet provided by the usual high level programming languages. They are available and can be used in PASCAL-XSC, [46, 47, 49, 67, 68], a PASCAL extension for the high accuracy scientific computing which was developed at the author's Institute. Roughly speaking, interval arithmetic brings guarantees into computation while the three scalar or dot products deliver high accuracy. These two features should not be confused.

The implementation of the 12 operations $\boxdot$, $\triangledown$, $\vartriangle$, $* \in \{+, -, \times, /\}$ on computers is routine and standard nowadays. Fast techniques are largely discussed in the literature. So we now turn to the implementation of the three optimal scalar products $\boxdot$, $\triangledown$ and $\vartriangle$ on computers. We shall discuss circuits for the hardware realization of these operations for different kinds of processors like personal computers, workstations, mainframes, super computers and digital signal processors. The differences in the circuits for these diverse processors are dictated by the speed with which the processor delivers the vector components a_i and b_i, $i = 1, 2, ..., n$, to the arithmetic or scalar product unit.

After a brief discussion of the implementation of the accurate scalar product on computers we shall detail two principal solutions to the problem. Solution A uses a long adder and a long shift. Solution B uses a short adder and some local memory in the arithmetic unit. At first sight both of these principal solutions seem to lead to relatively slow hardware circuits. However later, more refined studies will show that very fast circuits can be devised for both methods and for the diverse processors mentioned above. A first step in this direction is the provision of the very fast carry resolution scheme described in Section 1.2.4.

Actually it is a central result of this study that, for all processors under consideration, circuits for the computation of the optimal scalar product are available where virtually no computing time for the execution of the arithmetic is needed. In a pipeline, the arithmetic can be done within the time the processor needs to read the data into the arithmetic unit. This means that no other method to compute the scalar product can be faster, in particular, not even a conventional computation of scalar products in floating-point arithmetic which may lead to an incorrect answer. Once more we emphasize the fact that the methods to be discussed here compute the scalar product of two floating-point vectors of arbitrary finite length without loss of information or with only one final rounding at the very end of the computation.

Now we turn to our task. Let $a = (a_i)$ and $b = (b_i)$, $i = 1(1)n$, be two vectors with n components which are floating-point numbers, i. e.

$$a_i, b_i \in R(b, l, e1, e2), \text{ for } i = 1(1)n.$$

We are going to compute the two results (scalar products):

$$s := \sum_{i=1}^{n} a_i \times b_i = a_1 \times b_1 + a_2 \times b_2 + \ldots + a_n \times b_n,$$

and

$$c := \bigcirc \sum_{i=1}^{n} a_i \times b_i = \bigcirc(a_1 \times b_1 + a_2 \times b_2 + \ldots + a_n \times b_n) = \bigcirc s,$$

where all additions and multiplications are the operations for real numbers and $\bigcirc$ is a rounding symbol representing, for instance, rounding to nearest, rounding towards zero, rounding upwards or downwards.

Since a_i and b_i are floating-point numbers with a mantissa of l digits, the products $a_i \times b_i$ in the sums for s and c are floating-point numbers with a mantissa of $2l$ digits. The exponent range of these numbers doubles also, i. e. $a_i \times b_i \in R(b, 2l, 2e1, 2e2)$. All these summands can be expressed in a fixed-point register of length $2e2 + 2l + 2|e1|$ without loss of information, see Fig. 1.1. If one of the summands has an exponent 0, its mantissa can be expressed in a register of length $2l$. If another summand has exponent 1, it can be expressed with exponent 0, if the register provides further digits on the left and the mantissa is shifted one place to the left. An exponent -1 in one of the summands requires a corresponding shift to the right. The largest exponents in magnitude that may occur in the summands are $2e2$ and $2|e1|$. So all summands can be expressed with exponent 0 in a fixed-point register of length $2e2 + 2l + 2|e1|$ without loss of information.

1.2.1 Solution A: Long Adder and Long Shift

If the register is built as an accumulator with an adder, all summands could even be added without loss of information. In order to accommodate possible overflows, it is convenient to provide a few, say k more digits of base b on the left. In such an accumulator, every such sum or scalar product can be added without loss of information. As many as b^k overflows may occur and be accommodated for without loss of information. In the worst case, presuming every sum causes an overflow, we can accommodate sums with $n \leq b^k$ summands.

A gigaflops computer would perform about 10^{17} operations in 10 years. So 17 decimal or about 57 binary digits certainly are a reasonable upper bound for k. Thus, the long accumulator and the long adder consist of $L = k + 2e2 + 2l + 2|e1|$ digits of base b. The summands are shifted to the proper position and added. See Fig. 1.1. Fast carry resolution techniques will be discussed later. The final sums s and c are supposed to be in the single exponent range $e1 \leq e \leq e2$, otherwise c is not representable as a floating-point number and the problem has to be scaled.

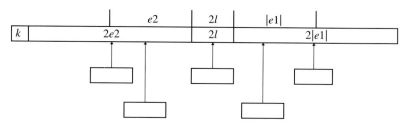

Fig. 1.1. Long accumulator with long shift for accurate scalar product accumulation.

1.2.2 Solution B: Short Adder with Local Memory on the Arithmetic Unit

In a scalar product computation the summands are all of length $2l$. So actually the long adder and long accumulator may be replaced by a short adder and a local store of size L on the arithmetic unit. The local store is organized in words of length l or l', where l' is a power of 2 and slightly larger than l. (For instance $l = 53$ bits and $l' = 64$ bits). Since the summands are of length $2l$, they fit into a part of the local store of length $3l'$. This part of the store is determined by the exponent of the summand. We load this part of the store into an accumulator of length $3l'$. The summand mantissa is placed in a shift register and is shifted to the correct position as determined by the exponent. Then the shift register contents are added to the contents of the accumulator. Fig. 1.2.

An addition into the accumulator may produce a carry. As a simple method to accommodate carries, we enlarge the accumulator on its left end by a few more digit positions. These positions are filled with the corresponding digits of the local store. If not all of these digits equal $b - 1$ in case of addition (or zero in case of subtraction), they will accommodate a possible carry of the addition (or borrow in case of subtraction). Of course, it is possible that all these additional digits are $b - 1$ (or zero). In this case, a loop can be provided that takes care of the carry and adds it to (subtracts it from) the next digits of the local store. This loop may need to be traversed several times. Other carry (borrow) handling processes are possible and will be dealt with later. This completes our sketch of the second method for an accurate computation of scalar products using a short adder and some local store on the arithmetic unit. See Fig. 1.2.

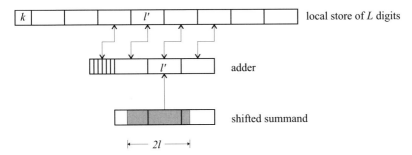

Fig. 1.2. Short adder and local store on the arithmetic unit for accurate scalar product accumulation

1.2.3 Remarks

The scalar product is a highly frequent operation in scientific computing. The two solutions A and B are both simple, straightforward and mature.

Remark 1: The purpose of the k digits on the left end of the register in Fig. 1.1 and Fig. 1.2 is to accommodate possible overflows. The only numbers that are added to this part of the register are plus or minus unity. So this part of the register just can be treated as a counter by an incrementer/decrementer.

Remark 2: The final result of a scalar product computation is assumed to be a floating-point number with an exponent in the range $e1 \leq e \leq e2$. During the computation, however, summands with an exponent outside of this range may very well occur. The remaining computation then has to cancel all these digits. This shows that normally in a scalar product computation, the register space outside the range $e1 \leq e \leq e2$ will be used less frequently. The conclusion should not be drawn from this consideration that the register size can be restricted to the single exponent range in order to save some silicon area. This would require the implementation of complicated exception handling routines which finally require as much silicon but do not solve the problem in principle.

Remark 3: We emphasize once more that the number of digits, L, needed for the register to compute scalar products of two vectors to full accuracy only depends on the floating-point data format. In particular it is independent of the number n of components of the two vectors to be multiplied.

As samples we calculate the register width L for a few typical and frequently used floating-point data formats:

a) IEEE-arithmetic single precision:
   $b = 2$; word length: 32 bits; sign: 1 bit; exponent: 8 bits; mantissa: $l = 24$ bits; exponent range: $e1 = -126$, $e2 = 127$, binary.
   $L = k + 2e2 + 2l + 2|e1| = k + 554$ bits.
   With $k = 86$ bits we obtain $L = 640$ bits. This register can be represented by 10 words of 64 bits.

b) /370 architecture, long data format:
   $b = 16$; word length: 64 bits; sign: 1 bit; mantissa: $l = 14$ hex digits; exponent range: $e1 = -64$, $e2 = 63$, hexadecimal.
   $L = k + 2e2 + 2l + 2|e1| = k + 282$ bits.
   With $k = 88$ bits we obtain $L = 88 + 4 * 282 = 1216$ bits. This register can be represented by 16 words of 64 bits.

c) IEEE-arithmetic double precision:
   $b = 2$; word length: 64 bits; sign: 1 bit; exponent: 11 bits; mantissa: $l = 53$ bits; exponent range: $e1 = -1022$, $e2 = 1023$, binary.
   $L = k + 2e2 + 2l + 2|e1| = k + 4196$ bits.
   With $k = 92$ bits we obtain $L = 4288$ bits. This register can be represented by 67 words of 64 bits.

These samples show that the register size (at a time where memory space is measured in gigabits and gigabytes) is modest in all cases. It grows with the exponent range of the data format. If this range should be extremely large, as for instance in case of an extended precision floating-point format, only an inner part of the register would be supported by hardware. The outer parts which then appear very rarely could be simulated in software. The long data format of the /370 architecture covers in decimal a range from about 10^{-75} to 10^{75} which is very modest. This architecture dominated the market for more than 20 years and most problems could conveniently be solved with machines of this architecture within this range of numbers.

Remark 4: Multiplication is often considered to be more complex than addition. In modern computer technology this is no longer the case. Very fast circuits for multiplication using carry-save-adders (Wallace tree) are available and common practice. They nearly equalize the time to compute a sum and a product of two floating-point numbers. In a scalar product computation usually a large number of products is to be computed. The multiplier is able to produce these products very quickly. In a balanced scalar product unit the accumulator should be able to absorb a product in about the same time the multiplier needs to produce it. Therefore, measures have to be taken to equalize the speed of both operations. Because of a possible long carry propagation the accumulation seems to be the more complicated process.

Remark 5: Techniques to implement the optimal scalar product on machines which do not provide enough register space on the arithmetic logical unit will be discussed in Section 1.6 later in this paper.

1.2.4 Fast Carry Resolution

Both solutions A and B for our problem which we sketched above seem to be slow at first glance. Solution A requires a long shift which is necessarily slow. The addition over perhaps 4000 bits is slow also, in particular if a long carry propagation is necessary. For solution B, five steps have to be carried out: 1. read from the local store, 2. perform the shift, 3. add the summand, 4. resolve the carry, possibly by loops, and 5. write the result back into the local store. Again the carry resolution may be very time consuming.

As a first step to speed up solutions A and B, we discuss a technique which allows a very fast carry resolution. Actually a possible carry can already be accommodated while the product, the addition of which might produce a carry, is still being computed.

Both solutions A and B require a long register in which the final sum in a scalar product computation is built up. Henceforth we shall call this register the *Long Accumulator* and abbreviate it as LA. It consists of L bits. LA is a fixed-point register wherein any sum of floating-point numbers and of simple products of floating-point numbers can be represented without error.

To be more specific we now assume that we are using the double precision data format of the IEEE-arithmetic standard 754. See case c) of remark 3. As soon as the principles are clear, a transfer of the technique to other data formats is easy. Thus, in particular, the mantissa consists of $l = 53$ bits. We assume additionally that the LA that appears in solutions A and B is subdivided into words of $l' = 64$ bits. The mantissa of the product $a_i \times b_i$ then is 106 bits wide. It touches at most three consecutive 64-bit words of the LA which are determined by the exponent of the product. A shifter then aligns the 106 bit product into the correct position for the subsequent addition into the three consecutive words of the LA. This addition may produce a carry (or a borrow in case of subtraction). The carry is absorbed by that next more significant 64 bit word of the LA in which not all digits are 1 (or 0 in case of subtraction). Fig. 1.3, a). For a fast detection of this word two information bits or flags are appended to each long accumulator word. Fig. 1.3, b). One of these bits, the *all bits 1* flag, is set to 1 if all 64 bits of the register word are 1. This means that a carry will propagate through the entire word. The other bit, the *all bits 0* flag, is set to 0, if all 64 bits of the register word are 0. This means that in case of subtraction a borrow will propagate through the entire word.

During the addition of a product into the three consecutive words of the LA, a search is started for the next more significant word of the LA where the *all bits 1* flag is not set. This is the word which will absorb a possible carry. If the addition generates a carry, this word must be incremented by one and all intermediate words must be changed from all bits 1 to all bits 0. The easiest way to do this is simply to switch the flag bits from *all bits 1* to *all bits 0* with the additional semantics that if a flag bit is set, the appropriate constant (all bits 0 or all bits 1) must be generated instead of reading the LA

18   1. Fast and Accurate Vector Operations

word contents when reading a LA word, Fig. 1.3, b). Borrows are handled in an analogous way.

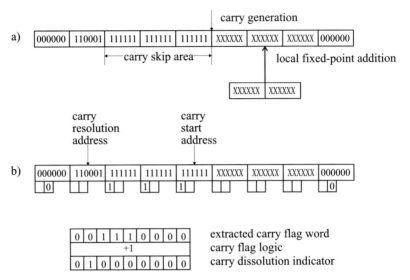

Fig. 1.3. Fast carry resolution

This carry handling scheme allows a very fast carry resolution. The generation of the carry resolution address is independent of the addition of the product, so it can be performed in parallel. At the same time, a second set of flags is set up for the case that a carry is generated. If the latter is the case, the carry is added into the appropriate word and the second set of flags is copied into the former flag word.

Simultaneously with the multiplication of the mantissa of a_i and b_i their exponents are added. This is just an eleven bit addition. The result is available very quickly. It delivers the exponent of the product and the address for its addition. By looking at the flags, the carry resolution address can be determined and the carry word can already be incremented/decremented as soon as the exponent of the product is available. It could be available before the multiplication of the mantissas is finished. If the accumulation of the product then produces a carry, the incremented/decremented carry word is written back into the LA, otherwise nothing is changed.

This very fast carry resolution technique could be used in particular for the computation of short scalar products which occur, for instance, in the computation of the real and imaginary part of a product of two complex floating-point numbers. A long scalar product, however, is usually performed in a pipeline. Then, during the execution of a product, the former product is added. It seems to be reasonable, then, to wait with the carry resolution until the former addition is actually finished.

1.3 High-Performance Scalar Product Units (SPU)

After having discussed the two principal Solutions A and B for exact scalar product computation as well as a very fast carry handling scheme, we now turn to a more detailed design of scalar product computation units for diverse processors. These units will be called SPU, which stands for Scalar Product Unit. If not otherwise mentioned we assume throughout this section that the data are stored in the double precision format of the IEEE-arithmetic standard 754. There the floating-point word has 64 bits and the mantissa consists of 53 bits. A central building block for the SPU is the long accumulator LA. It is a fixed-point register wherein any sum of floating-point numbers and of simple products of floating-point numbers can be represented without error. The unit allows the computation of scalar products of two vectors with any finite number of floating-point components to full accuracy or with one single rounding at the very end of the computation. As shown in Remark 3c) of Section 1.2.3, the LA consists of 4288 bits. It can be represented by 67 words of 64 bits.

The scalar product is a highly frequent operation in scientific computation. So its execution should be fast. All circuits to be discussed in this section perform the scalar product in a pipeline which simultaneously executes the following steps:

a) read the two factors a_i and b_i to perform a product,
b) compute the product $a_i \times b_i$ to the full double length, and
c) add the product $a_i \times b_i$ to the LA.

Step a) turns out to be the bottleneck of this pipeline. Therefore, we shall develop different circuits for computers which are able to read the two factors a_i and b_i into the SPU in four or two or one portion. The latter case will be discussed in Section 1.5. Step b) produces a product of 106 bits. It maps onto at most three consecutive words of the LA. The address of these words is determined by the products exponent. In step c) the 106 bit product is added to the three consecutive words of the LA.

1.3.1 SPU for Computers with a 32 Bit Data Bus

Here we consider a computer which is able to read the data into the arithmetic logical unit and/or the SPU in portions of 32 bits. The personal computer is a typical representative of this kind of computer.

Solution A with an adder and a shifter for the full LA of 4288 bits would be too expensive. So the SPU for these computers is built upon solution B

20   1. Fast and Accurate Vector Operations

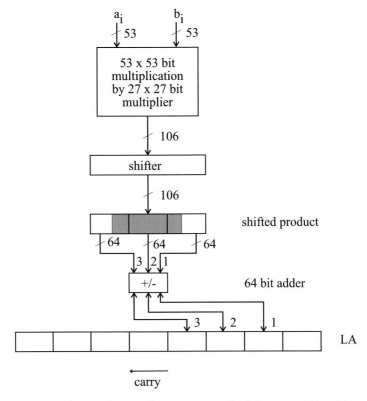

Fig. 1.4. Accumulation of a product to the LA by a 64 bit adder

(see Fig. 1.4). For the computation of the product $a_i \times b_i$ the two factors a_i and b_i are to be read. Both consist of 64 bits. Since the data can only be read in 32 bit portions, the unit has to read 4 times. We assume that with the necessary decoding this can be done in eight cycles. See Fig. 1.5. This is rather slow and turns out to be the bottleneck for the whole pipeline. In a balanced SPU the multiplier should be able to produce a product and the adder should be able to accumulate the product in about the same time the unit needs to read the data. Therefore, it suffices to provide a 27×27 bit multiplier. It computes the 106 bit product of the two 53 bit mantissas of a_i and b_i by 4 partial products. The subsequent addition of the product into the three consecutive words of the LA is performed by an adder of 64 bits. The appropriate three words of the LA are loaded into the adder one after the other and the appropriate portion of the product is added. The sum is written back into the same word of the LA where the portion has been read from. A 64 out of 106 bit shifter must be used to align the product onto the relevant word boundaries. See Fig. 1.4. The addition of the three portions of the product into the LA may cause a carry. The carry is absorbed

1.3 High-Performance Scalar Product Units (SPU)

cycle	read	mult/shift	accumulate				
	read a^1_{i-1}						
	read a^2_{i-1}						
	read b^1_{i-1}						
	read b^2_{i-1}						
	read a^1_i						
	read a^2_i	$c_{i-1} := a_{i-1} * b_{i-1}$					
	read b^1_i	$c_{i-1} := \text{shift}(c_{i-1})$					
	read b^2_i						
	read a^1_{i+1}		load1				
			add/sub	load2			
	read a^2_{i+1}	$c_i := a_i * b_i$	store1	add/sub	load3		
				store2	add/sub	load carry	
	read b^1_{i+1}	$c_i := \text{shift}(c_i)$			store3	inc/dec	
						store carry	
	read b^2_{i+1}					store flags	
	read a^1_{i+2}		load1				
			add/sub	load2			
	read a^2_{i+2}	$c_{i+1} := a_{i+1} * b_{i+1}$	store1	add/sub	load3		
				store2	add/sub	load carry	
	read b^1_{i+2}	$c_{i+1} := \text{shift}(c_{i+1})$			store3	inc/dec	
						store carry	
	read b^2_{i+2}					store flags	
	read a^1_{i+3}		load1				
			add/sub	load2			
	read a^2_{i+3}	$c_{i+2} := a_{i+2} * b_{i+2}$	store1	add/sub	load3		
				store2	add/sub	load carry	
	read b^1_{i+3}	$c_{i+2} := \text{shift}(c_{i+2})$			store3	inc/dec	
						store carry	
	read b^2_{i+3}					store flags	

Fig. 1.5. Pipeline for the accumulation of scalar products on computers with 32 bit data bus.

by incrementing (or decrementing in case of subtraction) a more significant word of the LA as determined by the carry handling scheme.

A brief sketch of the pipeline is shown in Fig. 1.5. There, we assume that a dual port RAM is available on the SPU to store the LA. This is usual for register memory. It allows simultaneous reading from the LA and writing into the LA. Eight machine cycles are needed to read the two 64 bit factors a_i and b_i for a product, including the necessary decoding of the data. This is also about the time in which the multiplication and the shift can be performed in the second step of the pipeline. The three successive additions and the carry resolution in the third step of the pipeline again can be done in about the

22   1. Fast and Accurate Vector Operations

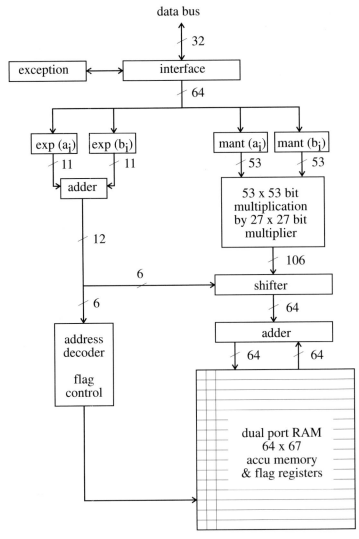

Fig. 1.6. Block diagram for a SPU with 32 bit data supply and sequential addition into SPU

same time. See Fig. 1.5. Fig. 1.6 shows a block diagram for a SPU with 32 bit data bus.

The sum of the exponents of a_i and b_i delivers the exponent of the product $a_i \times b_i$. It consists of 12 bits. The 6 low order (less significant) bits of this sum are used to perform the shift. The more significant bits of the sum deliver the LA address to which the product $a_i \times b_i$ has to be added. So the originally very long shift is split into a short shift and an addressing operation. The shifter performs a relatively short shift operation. The addressing selects the three words of the LA for the addition of the product.

The LA RAM needs only one address decoder to find the start address for an addition. The two more significant parts of the product are added to the contents of the two LA words with the two subsequent addresses. The carry logic determines the word which absorbs the carry. All these address decodings can be hard wired. The result of each one of the four additions is written back into the same LA word to which the addition has been executed. The two carry flags appended to each accumulator word are indicated in Fig. 1.6. In practice the flags are kept in separate registers.

We stress the fact that in the circuit just discussed virtually no specific computing time is needed for the execution of the arithmetic. In the pipeline the arithmetic is performed in the time which is needed to read the data into the SPU. Here, we assumed that this requires 8 cycles. This allows both the multiplication and the accumulation to be performed very economically and sequentially by a 27×27 bit multiplier and a 64 bit adder. Both the multiplication and the addition are themselves performed in a pipeline. The arithmetic overlaps with the loading of the data into the SPU.

There are processors on the market, where the data supply to the arithmetic unit or the SPU is much faster. We discuss the design of a SPU for such processors in the next section and in Section 1.5.

1.3.2 SPU for Computers with a 64 Bit Data Bus

Now we consider a computer which is able to read the data into the arithmetic logical unit and/or the SPU in portions of 64 bits. Fast workstations or mainframes are typical for this kind of computer.

Now the time to perform the multiplication and the accumulation overlapped in pipelines as before is no longer available. In order to keep the execution time for the arithmetic within the time the SPU needs to read the data, we have to invest in more hardware. For the multiplication a 53×53 bit multiplier must now be used. The result is still 106 bits wide. It could touch three 64 bit words of the LA. But the addition of the product and the carry resolution now have to be performed in parallel.

The 106 bit summand may fit into two instead of three consecutive 64 bit words of the LA. A closer look at the details shows that the 22 least significant bits of the three consecutive LA words are never changed by an addition of the 106 bit product. Thus the adder needs to be 170 bits wide only. Fig. 1.7 shows a sketch for the parallel accumulation of a product.

In the circuit a 106 to 170 bit shifter is used. The four additions are to be performed in parallel. So four read/write ports are to be provided for the LA RAM. A sophisticated logic must be used for the generation of the carry resolution address, since this address must be generated very quickly. Again the LA RAM needs only one address decoder to find the start address for an addition. The more significant parts of the product are added to the contents

24   1. Fast and Accurate Vector Operations

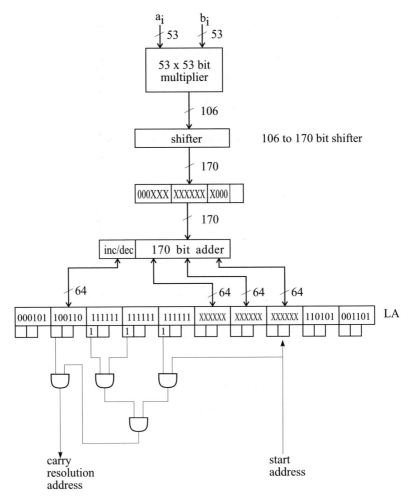

Fig. 1.7. Parallel accumulation of a product into the LA

of the two LA words with the two subsequent addresses. A tree structured carry logic now determines the LA word which absorbs the carry. A very fast hardwired multi-port driver can be designed which allows all 4 LA words to be read into the adder in one cycle.

Fig. 1.8 shows the pipeline for this kind of addition. In the figure we assume that 2 machine cycles are needed to decode and read one 64 bit word into the SPU.

Fig. 1.9 shows a block diagram for a SPU with a 64 bit data bus and parallel addition.

We emphasize again that virtually no computing time is needed for the execution of the arithmetic. In a pipeline the arithmetic is performed in the

cycle	read	mult/shift	accumulate
	read a_{i-1}		
	read b_{i-1}		
	read a_i	$c_{i-1} := a_{i-1} * b_{i-1}$	
	read b_i	$c_{i-1} := $ shift (c_{i-1})	
	read a_{i+1}	$c_i := a_i * b_i$	address decoding load
	read b_{i+1}	$c_i := $ shift (c_i)	add/sub c_{i-1} store & store flags
	read a_{i+2}	$c_{i+1} := a_{i+1} * b_{i+1}$	address decoding load
	read b_{i+2}	$c_{i+1} := $ shift (c_{i+1})	add/sub c_i store & store flags
	read a_{i+3}	$c_{i+2} := a_{i+2} * b_{i+2}$	address decoding load
	read b_{i+3}	$c_{i+2} := $ shift (c_{i+2})	add/sub c_{i+1} store & store flags

Fig. 1.8. Pipeline for the accumulation of scalar products.

time which is needed to read the data into the SPU. Here, we assume that with the necessary decoding, this requires 4 cycles for the two 64 bit factors a_i and b_i for a product. To match the shorter time required to read the data, more hardware has to be invested for the multiplier and the adder.

If the technology is fast enough it may be reasonable to provide a 256 bit adder instead of the 170 bit adder. An adder width of a power of 2 may simplify the shift operation as well as the address decoding. The lower bits of the exponent of the product control the shift operation while the higher bits are directly used as the start address for the accumulation of the product into the LA.

The two flag registers appended to each accumulator word are indicated in Fig. 1.9 again. In practice the flags are kept in separate registers.

1.4 Comments on the Scalar Product Units

1.4.1 Rounding

If the result of an exact scalar product is needed later in a program, the contents of the LA must be put into the user memory. How this can be done will be discussed later in this section.

26   1. Fast and Accurate Vector Operations

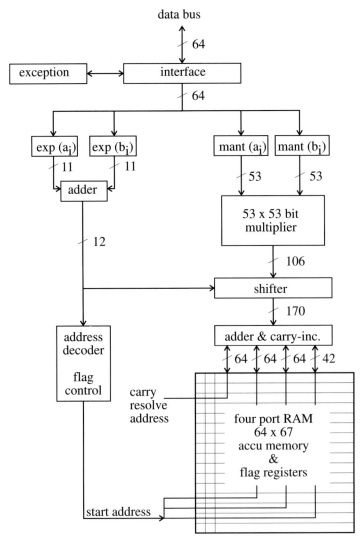

Fig. 1.9. Block diagram for a SPU with 64 bit data bus and parallel addition into the SPU

If not processed any further the correct result of a scalar product computation usually has to be rounded into a floating-point number or a floating-point interval. The flag bits that are used for the fast carry resolution can be used for the rounding of the LA contents also. By looking at the flag bits, the leading result word in the accumulator can easily be identified. This and the next LA word are needed to compose the mantissa of the result. This 128 bit quantity must then be shifted to form a normalized mantissa of an IEEE-arithmetic double precision number. The shift length can be extracted by looking at the leading result word in the accumulator with the same procedure which identified it by looking at the flag bit word.

For the correct execution of the rounding downwards (or upwards) it is necessary to check whether any one of the discarded bits is different from zero. This is done by testing the remaining bits of the 128 bit quantity in the shifter and by looking at the *all bits 0* flags of the following LA words. This information then is used to control the rounding.

Only one rounding at the very end of a scalar product computation is needed. If a large number of products has been accumulated the contribution of the rounding to the computing time is not substantial. However, if a short scalar products or a single floating-point operation addition or subtraction has to be carried out by the SPU, a very fast rounding procedure is essential for the speed of the overall operation.

The rounding depends heavily on the speed with which the leading non zero digit of the LA can be detected. A pointer to this digit, carried along with the computation, would immediately identify this digit. The pointer logic requires additional hardware and its usefulness decreases for lengthy scalar products to be computed.

For short scalar products or single floating-point operations leading zero anticipation (LZA) would be more useful. The final result of a scalar product computation is supposed to lie in the exponent range between $e1$ and $e2$ of the LA. Otherwise the problem has to be scaled. So hardware support for the LZA is only needed for this part of the LA. A comparison of the exponents of the summands identifies the LA word for which the LZA should be activated. The LZA consists of a fast computation of a provisional sum which differs from the correct sum by at most one leading zero. With this information the leading zeros and the shift width for the two LA words in question can be detected easily and fast. [91].

1.4.2 How much Local Memory should be Provided on a SPU?

There are applications which make it desirable to provide more than one long accumulator on the SPU. If, for instance, the components of the two vectors $a = (a_i)$ and $b = (b_i)$ are complex floating-point numbers, the scalar product $a \cdot b$ is also a complex floating-point number. It is obtained by accumulating the real and imaginary parts of the product of two complex floating-point numbers. The formula for the product of two complex floating-point numbers

$$(x = x_1 + ix_2,\ y = y_1 + iy_2) \Rightarrow$$
$$x \times y = (x_1 \times y_1 - x_2 \times y_2) + i\,(x_1 \times y_2 + x_2 \times y_1))$$

shows that the real and imaginary part of a_i and b_i are needed for the computation of both the real part of the product $a_i \times b_i$ as well as the imaginary part.

Access to user memory is usually slower than access to register memory. To obtain high computing speed it is desirable, therefore, to read the real and imaginary parts of the vector components only once and to compute the real and imaginary parts of the products simultaneously in two long accumulators on the SPU instead of reading the data twice and performing the two accumulations sequentially. The old calculators shown in Fig. 1.22 and Fig. 1.23 on page 62 had already two long registers.

Very similar considerations show that a high speed computation of the scalar product of two vectors with interval components makes two long accumulators desirable as well.

There might be other reasons to provide local memory space for more than one LA on the SPU. A program with higher priority may interrupt the computation of a scalar product and require a LA. The easiest way to solve this problem is to open a new LA for the program with higher priority. Of course, this can happen several times which raises the question how much local memory for how many long accumulators should be provided on a SPU. Three might be a good number to solve this problem. If a further interrupt requires another LA, the LA with the lowest priority could be mapped into the main memory by some kind of stack mechanism and so on. This technique would not limit the number of interrupts that may occur during a scalar product computation. These problems and questions must be solved in connection with the operating system.

For a time sharing environment memory space for more than one LA on the SPU may also be useful.

However the contents of the last two paragraphs are of a more hypothetical nature. The author is of the opinion that the scalar product is a fundamental and basic operation which should not and never needs to be interrupted.

1.4.3 A SPU Instruction Set

For the SPU the following 10 instructions for the LA are recommended. These 10 low level instructions are most natural and inevitable as soon as the idea of the long accumulator for the accurate scalar product has been chosen. They are low level capabilities to support the high level instructions developed in the next section, and are based on preceding experience with these in the XSC-languages since 1980. Very similar instructions were provided by the processor developed in [93]. Practically identical instructions were used in [109] to support ACRITH and ACRITH-XSC [108, 110, 112]. These IBM program products have been developed at the author's institute in collaboration with IBM.

The 10 low level instructions for the LA are:

1. clear the LA,
2. add a product to the LA,
3. add a floating-point number to the LA,

1.4 Comments on the Scalar Product Units

4. subtract a product from the LA,
5. subtract a floating-point number from the LA,
6. read LA and round to the destination format,
7. store LA contents in memory,
8. load LA contents from memory,
9. add LA to LA,
10. Subtract LA from LA.

The clear instruction can be performed by setting all *all bits 0* flags to 0. The load and store instructions are performed by using the load/store instructions of the processor. For the add, subtract and round instructions the following denotations could be used. There the prefix *sp* identifies SPU instructions. *ln* denotes the floating-point format that is used and will be *db* for IEEE double. In all SPU instructions, the LA is an implicit source and destination operand. The number of the instruction above is repeated in the following coding which could be used to realize it.

2. spadd *ln* src1, src2,
   multiply the numbers in the given registers and add the product to the LA.
3. spadd *ln* src,
   add the number in the given register to the LA.
4. spsub *ln* src1, src2,
   multiply the numbers in the given registers and subtract the product from the LA.
5. spsub *ln* src,
   subtract the number in the given register from the LA.
6. spstore *ln.rd* dest,
   get LA contents and put the rounded value into the destination register. In the instruction *rd* controls the rounding mode that is used when the LA contents is stored in a floating-point register. It is one of the following:
   - *rn*   round to nearest,
   - *rz*   round towards zero,
   - *rp*   round upwards, i. e. towards plus infinity,
   - *rm*   round downwards, i. e. towards minus infinity.
7. spstore dest,
   get LA contents and put its value into the destination memory operand.
8. spload src,
   load accumulator contents from the given memory operand into the LA.
9. spadd src,
   the contents of the accumulator at the location src are added to the contents of the accumulator in the processor.
10. spsub src,
    the contents of the accumulator at the location src are subtracted from the contents of the accumulator in the processor.

1.4.4 Interaction with High Level Programming Languages

This paper is motivated by the tremendous advances in computer technology that have been made in recent years. 100 million transistors can be placed on a single chip. This allows the quality and high accuracy of the basic floating-point operations of addition, subtraction, multiplication and division to be extended to the arithmetic operations in the linear spaces and their interval extensions which are most commonly used in computation. A new fundamental operation, the scalar product, is needed to provide this advanced computer arithmetic. The scalar product can be produced by an instruction *multiply and accumulate* and placed in the LA which has enough digit positions to contain the exact sum without rounding. Only a single rounding error of at most one unit in the last place is introduced when the completed scalar product (often also called dot product) is returned to one of the floating-point registers.

By operator overloading in modern programming languages matrix and vector operations can be provided with highest accuracy and in a simple notation, if the optimal scalar product is available. However, many scalar products that occur in a computation do not appear as vector or matrix operations in the program. A vectorizing compiler is certainly a good tool detecting such additional scalar products in a program. Since the hardware supported optimal scalar product is faster than a conventional computation in floating-point arithmetic this would increase both the accuracy and the speed of the computation.

In the computer, the scalar product is produced by several, more elementary computer instructions as shown in the last section. Programming and the detection of scalar products in a program can be simplified a great deal if several of these computer instructions are put into the hands of the user and incorporated into high level programming languages. This has been done with great success in the so-called XSC-languages (eXtended Scientific Computation) since 1980 [14, 41, 46–49, 56, 67, 68, 106–108, 112] that have been developed at the author's institute. All these languages provide an accurate scalar product implemented in software based on integer arithmetic. If a computer is equipped with the hardware unit XPA 3233 (see Section 1.7) the hardware unit is called instead. A large number of problem solving routines with automatic result verification has been implemented in the XSC-languages for practically all standard problems of numerical analysis [34, 35, 57, 64, 110, 112]. These routines have very successfully been applied in the sciences.

We mention a few of these constructs and demonstrate their usefulness. Central to this is the idea of allowing variables of the size of the LA to be defined in a user's program. For this purpose a new data type called *dotprecision* is introduced. A variable of the type *dotprecision* is a fixed-point variable with $L = k + 2e2 + 2l + 2|e1|$ digits of base b. See Fig. 1.1. As has been shown earlier, every finite sum of floating-point products $\sum_{i=1}^{n} a_i \times b_i$

1.4 Comments on the Scalar Product Units

can be represented as a variable of type *dotprecision*. Moreover, every such sum can be computed in a local store of length L on the SPU without loss of information. Along with the type *dotprecision* the following constructs serve as primitives for developing expressions in a program which can easily be evaluated with the SPU instruction set:

dotprecision   new data type
:=   assignment from dotprecision
                     to dotprecision or
                     to real with rounding to nearest or
                     to interval with roundings downwards and upwards
                     depending on the type on the left hand side of the
                     := operator.

For variables of type dotprecision so-called *dotprecision expressions* are permitted which are defined as sums of *simple expressions*. A *simple expression* is either a signed or unsigned constant or a variable of type real or a single product of two such objects or another dotprecision variable. All operations (multiplications and accumulations) are to be executed to full accuracy.

For instance, let x be a variable of type dotprecision and y and z variables of type real. Then in the assignment

x := x + y * z

the double length product of y and z is added to the variable x of type dotprecision and its new value is assigned to x.

The scalar product of two vectors $a = (a_i)$ and $b = (b_i)$ is now easily implemented with a variable x of type dotprecision as follows:

x := 0;
for i := 1 **to** n **do** x := x + a[i] * b[i];
y := x;

The last statement $y := x$ rounds the value of the variable x of type dotprecision into the variable y of type real by applying the standard rounding of the computer. y then has the value of the scalar product $a \square b$ which is within a single rounding error of the exact scalar product $a \cdot b$.

For example, the method of defect correction or iterative refinement requires highly accurate computation of expressions of the form

$$a \cdot b - c \cdot d$$

with vectors a, b, c, and d. Employing a variable x of type dotprecision, this expression can now be programmed as follows:

x := 0;
for i := 1 **to** n **do** x := x + a[i] * b[i];
for i := 1 **to** n **do** x := x − c[i] * d[i];
y := x;

32   1. Fast and Accurate Vector Operations

The result, involving $2n$ multiplications and $2n - 1$ additions, is produced with but a single rounding operation.

In the last two examples y could have been defined to be of type interval. Then the last statement $y := x$ would produce an interval with a lower bound which is obtained by rounding the dotprecision value of x downwards and an upper bound by rounding it upwards. Thus, the bounds of y will be either the same or two adjacent floating-point numbers.

In the XSC-languages the functionality of the dotprecision type and expression is available also for complex data as well as for interval and complex interval data.

1.5 Scalar Product Units for Top-Performance Computers

By definition a top-performance computer is able to read two data x and y to perform a product $x \times y$ into the arithmetic logical unit and/or the SPU simultaneously in one portion. Supercomputers and vector processors are typical representatives of this kind of computers. Usually the floating-point word consists of 64 bits and the data bus is 128 or even more bits wide. However, digital signal processors with a word size of 32 bits can also belong in this class if two 32 bit words are read into the ALU and/or SPU in one portion. For these kind of computers both solutions A and B which have been sketched in Sections 1.2.1 and 1.2.2 make sense and will be discussed. The higher the speed of the system the more hardware has to be employed. The most involved and expensive solution seems to be best suited to reveal the basic ideas. So we begin with solution A using a long adder for the double precision data format.

1.5.1 Long Adder for 64 Bit Data Word (Solution A)

In [44] the basic ideas have been developed for a general data format. However, to be very specific we discuss here a circuit for the double precision format of the IEEE-arithmetic standard 754. The word size is 64 bits. The mantissa has 53 bits and the exponent 11 bits. The exponent covers a range from -1022 to $+1023$. The LA has 4288 bits. We assume again that the scalar product computation can be subdivided into a number of independent steps like

   a) read a_i and b_i,
   b) compute the product $a_i \times b_i$,
   c) add the product to the LA.

Now by assumption the SPU can read the two factors a_i and b_i simultaneously in one portion. We call the time that is needed for this a *cycle*.

1.5 Scalar Product Units for Top-Performance Computers

Then, in a balanced design, steps b) and c) should both be performed in about the same time. Using well known fast multiplication techniques like Booth-Recoding and Wallace-tree this certainly is possible for step b). Here, the two 53 bit mantissas are multiplied. The product has 106 bits. The main difficulty seems to appear in step c). There, we have to add a summand of 106 bits to the LA in every *cycle*.

With solution A the addition is performed by a long adder and a long shift, both of $L = 4288$ bits. An adder and a shift of this size are necessarily slow, certainly too slow to process one summand of 106 bits in a single *cycle*. Therefore, measures have to be taken to speed up the addition as well as the shift. As a first step we subdivide the long adder into shorter segments. Without loss of generality we assume that the segments consist of 64 bits.[4] A 64 bit adder certainly is faster than a 4288 bit adder. Now each one of the 64 bit adders may produce a carry. We write these carries into carry registers between two adjacent adders. See Fig. 1.10.

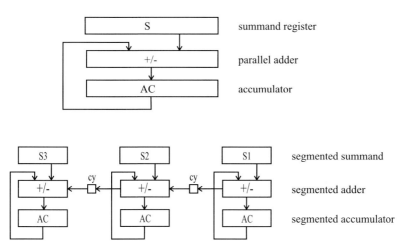

Fig. 1.10. Parallel and segmented parallel adder

If a single addition has to be performed these carries still have to be propagated. In a scalar product computation, however, this is not necessary. We assume that a large number of summands has to be added. We simply add the carry with the next summand to the next more significant adder. Only at the very end of the accumulation, when no more summands are coming, carries may have to be eliminated. However, every summand is relatively short. It consists of 106 bits only. So during the addition of a summand, carries are only produced in a small part of the 4288 bit adder. The carry elimination, on the other hand, takes place during each step of the addition

---
[4] other segments are possible, see [44, 45].

wherever a carry is left. So in an average case there will only be very few carries left at the end of the accumulation and a few additional *cycles* will suffice to absorb the remaining carries. Thus, segmenting the adder enables it to keep up in speed with steps a) and b) and to read and process a summand in each *cycle*.

The long shift of the 106 bit summand is slow also. It is speeded up by a matrix shaped arrangement of the adders. Only a few, let us assume here four of the partial adders, are placed in a row. We begin with the four least significant adders. The four next more significant adders are placed directly beneath of them and so on. The most significant adders form the last row. The rows are connected as shown in Fig. 1.11.

In our example, where we have 67 adders of 64 bits, 17 rows suffice to arrange the entire summing matrix. Now the long shift is performed as follows: The summand of 106 bits carries an exponent. In a fast shifter of 106 to 256 bits it is shifted into a position where its most significant digit is placed directly above the position in the long adder which carries the same exponent identification E. The remaining digits of the summand are placed immediately to its right. Now the summing matrix reads this summand into the S-registers (summand registers) of every row. The addition is executed in that row where the exponent identification coincides with that of the summand.

It may happen that the most significant digit of the summand has to be shifted so far to the right that the remaining digits would hang over at the right end of the shifter. These digits then are reinserted at the left end of the shifter by a ring shift. If now the more significant part of the summand is added in row r, its less significant part will be added in row $r - 1$.

By this matrix shaped arrangement of the adders, the unit can perform both a shift and an addition in a single *cycle*. The long shift is reduced to a short shift of 106 to 256 bits which is fast. The remaining shift happens automatically by the row selection for the addition in the summing matrix.

Every summand carries an exponent which in our example consists of 12 bits. The lower part of the exponent, i. e. the 8 least significant digits, determine the shift width and with it the selection of the columns in the summing matrix. The row selection is obtained by the 4 most significant bits of the exponent. This complies roughly with the selection of the adding position in two steps by the process of Fig. 1.2. The shift width and the row selection for the addition of a product $a_i \times b_i$ to the LA are known as soon as the exponent of the product has been computed. Since the exponents of a_i and b_i consist of 11 bits only, the result of their addition is available very quickly. So while the multiplication of the mantissas is still being executed the shifter can already be switched and the addresses of the LA words for the accumulation of the product $a_i \times b_i$ can be selected.

The 106 bit summand touches at most three consecutive words of the LA. The addition of the summand is executed by these three partial adders.

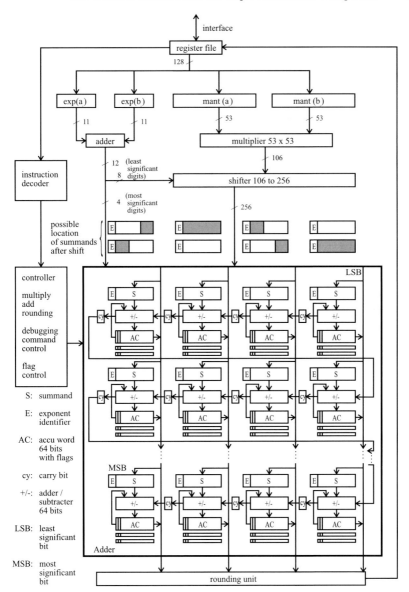

Fig. 1.11. Block diagram of a SPU with long adder for a 64 bit data word and 128 bit data bus

Each of these adders can produce a carry. The carry of the leftmost of these partial adders can with high probability be absorbed, if the addition always is executed over four adders and the fourth adder then is the next more significant one. This can reduce the number of carries that have to be resolved during future steps of the accumulation and in particular at the end.

In each step of the accumulation an addition only has to be activated in the selected row of adders and in those adders where a non zero carry is waiting to be absorbed. This adder selection can reduce the power consumption for the accumulation step significantly.

The carry resolution method that has been discussed so far is quite natural. It is simple and does not require particular hardware support. If long scalar products are being computed it works very well. Only at the end of the accumulation, if no more summands are coming, a few additional *cycles* may be required to absorb the remaining carries. Then a rounding can be executed. However, this number of additional *cycles* for the carry resolution at the end of the accumulation, although it is small in general, depends on the data and is unpredictable. In case of short scalar products the time needed for these additional *cycles* may be disproportionately high and indeed exceed the addition time.

With the fast carry resolution mechanism that has been discussed in Section 1.2.4 these difficulties can be overcome. At the cost of some additional hardware all carries can be absorbed immediately at each step of the accumulation. The method is shown in Fig. 1.11 also. Two flag registers for the *all bits 0* and the *all bits 1* flags are shown at the left end of each partial accumulator word in the figure. The addition of the 106 bit products is executed by three consecutive partial adders. Each one of these adders can produce a carry. The carries between two of these adjacent adders can be avoided, if all partial adders are built as *Carry Select Adders*. This increases the hardware costs only moderately. The carry registers between two adjacent adders then are no longer necessary.[5] The flags indicate which one of the more significant LA words will absorb the left most carry. During an addition of a product only these 4 LA words are changed and only these 4 adders need to be activated. The addresses of these 4 words are available as soon as the exponent of the summand $a_i \times b_i$ has been computed. During the addition step now simultaneously with the addition of the product the carry word can be incremented (decremented). If the addition produces a carry the incremented word will be written back into the local accumulator. If the addition does not produce a carry, the local accumulator word remains unchanged. Since we have assumed that all partial adders are built as *Carry Select Adders* this final carry resolution scheme requires no additional hardware. Simultaneously with the incrementation/decrementation of the carry word a second set of

---

[5] This is the case in Fig. 1.12 where a similar situation is discussed. There all adders are supposed to be carry select adders.

flags is set up for the case that a carry is generated. In this case the second set of flags is copied into the former word.

The accumulators that belong to partial adders in Fig. 1.11 are denoted by AC. Beneath them a small memory is indicated in the figure. It can be used to save the LA contents very quickly in case that a program with higher priority interrupts the computation of a scalar product and requires the unit for itself. However, the author is of the opinion that the scalar product is a fundamental and basic arithmetic operation which should never be interrupted. The local memory on the SPU can be used for fast execution of scalar products in the case of complex and of interval arithmetic.

In Section 1.4.2 we have discussed applications like complex arithmetic or interval arithmetic which make it desirable to provide more than one LA on the SPU. The local memory on the SPU shown in Fig. 1.11 serves this purpose.

In Fig. 1.11 the registers for the summands carry an exponent identification denoted by E. This is very useful for the final rounding. The usefulness of the flags for the final rounding has already been discussed. They also serve for fast clearing of the accumulator.

The SPU which has been discussed in this section seems to be costly. However, it consists of a large number of identical parts and it is very regular. This allows a highly compact design. Furthermore the entire unit is simple. No particular exception handling techniques are to be dealt with by the hardware. Vector computers are the most expensive. A compact and simple solution, though expensive, is justified for these systems.

1.5.2 Long Adder for 32 Bit Data Word (Solution A)

In this section as well as in Section 1.5.4 we consider a computer which uses a 32 bit floating-point word and which is able to read two such words into the ALU and/or SPU simultaneously in one portion. Digital signal processors are representatives of this kind of computer. Real time computing requires very high computing speed and high accuracy in the result. As in the last section we call the time that is needed to read the two 32 bit floating-point words a *cycle*.

We first develop circuitry which realizes Solution A using a long adder and a long shift. To be very specific we assume that the data are given as single precision floating-point numbers conforming to the IEEE-arithmetic standards 754. There the mantissa consists of 24 bits and the exponent has 8 bits. The exponent covers a range from -126 to $+127$ (in binary). As discussed in Remark 3a) of Section 1.2.3, 640 bits are a reasonable choice for the LA. It can be represented by 10 words of 64 bits.

Again the scalar product is computed by a number of independent steps like

a) read a_i and b_i,
b) compute the product $a_i \times b_i$,
c) add the product to the LA.

Each of the mantissas of a_i and b_i has 24 bits. Their product has 48 bits. It can be computed very fast by a 24×24 bit multiplier using standard techniques like Booth-Recoding and Wallace tree. The addition of the two 8 bit exponents of a_i and b_i delivers the exponent of the product consisting of 9 bits.

The LA consists of 10 words of 64 bits. The 48 bit mantissa of the product touches at most two of these words. The addition of the product is executed by the corresponding two consecutive partial adders. Each of these two adders can produce a carry. The carry between the two adjacent adders can immediately be absorbed if all partial adders are built as *Carry Select Adders* again. The carry of the more significant of the two adders will be absorbed by one of the more significant 64 bit words of the LA. The flag mechanism (see Section 1.2.4) indicates which one of the LA words will absorb a possible carry. So during an addition of a summand the contents of at most 3 LA words are changed and only these three partial adders need to be activated. The addresses of these words are available as soon as the exponent of the summand $a_i \times b_i$ has been computed. During the addition step, simultaneously with the addition of the product, the carry word can be incremented (decremented). If the addition produces a carry the incremented word will be written back into the local accumulator. If the addition does not produce a carry, the local accumulator word remains unchanged. Since all partial adders are built as *Carry Select Adders* no additional hardware is needed for the carry resolution. Simultaneously with the incrementation/decrementation of the carry word a second set of flags is set up for the case that a carry is generated. If the latter is the case the second set of flags is copied into the former flag word.

Details of the circuitry just discussed are summarized in Fig. 1.12. The figure is highly similar to Fig. 1.11 of the previous section. In order to avoid the long shift, the long adder is designed as a summing matrix consisting of 2 adders of 64 bits in each row. For simplicity in the figure only 3 rows (of the 5 needed to represent the full LA) are shown.

In a fast shifter of 48 to 128 bits the 48 bit product is shifted into a position where its most significant digit is placed directly above the position in the long adder which carries the same exponent identification E. The remaining digits of the summand are placed immediately to its right. If they hang over at the right end of the shifter, they are reinserted at the left end by a ring shift. Above the summing matrix in Fig. 1.12 two possible positions of summands after the shift are indicated.

The summing matrix now reads the summand into its S-registers. The addition is executed by those adders where the exponent identification coincides with the one of the summand. The exponent of the summand consists

1.5 Scalar Product Units for Top-Performance Computers   39

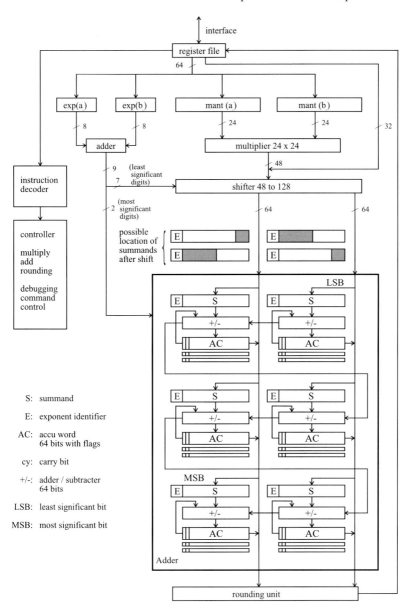

Fig. 1.12. Block diagram of a SPU with long adder for a 32 bit data word and 64 bit data bus

of 9 bits. The lower part, i. e. the 7 least significant digits, determine the shift width. The selection of the two adders which perform the addition is determined by the most significant bits of the exponent.

In Fig. 1.12 again some memory is indicated for each part of the LA. It can be used to save the LA contents very quickly in case a program with higher priority interrupts the computation of a scalar product and requires the unit for itself. The local memory on the SPU also can be used for fast execution of scalar products in the case of complex arithmetic and of interval arithmetic.

In comparison with Fig. 1.11, Fig. 1.12 shows an additional 32 bit data path directly from the input register file to the fast shifter. This data path is supposed to allow a very fast execution of the operation *multiply and add fused*, $rnd(a \times b + c)$, which is provided by some conventional floating-point processors. While the product $a \times b$ is computed by the multiplier, the summand c is added to the LA.

The SPU which has been discussed in this section seems to be costly at first glance. While a single floating-point addition conveniently can be done with one 64 bit adder, here 640 full adders (10 64-bit adders) have been used in carry select adder mode. However, the advantages of this design are tremendous. While a conventional floating-point addition can produce a completely wrong result with only two or three additions, the new unit never delivers a wrong answer, even if millions of floating-point numbers or single products of such numbers are added. An error analysis is never necessary for these operations. The unit consists of a large number of identical parts and it is very regular. This allows a very compact design. No particular hardware has to be included to deal with rare exceptions. Although an increase in adder equipment by a factor of 10, compared with a conventional floating-point adder, might seem to be high, the number of full adders used for the circuitry is not extraordinary. We stress the fact that for a Wallace tree in case of a standard 53×53 bit multiplier about the same number of full adders is used. For fast conventional computers this has been the state of the art multiplication for many years and nobody complains about high cost.

1.5.3 Short Adder with Local Memory on the Arithmetic Unit for 64 Bit Data Word (Solution B)

In the circuits discussed in Sections 1.5.1 and 1.5.2 adder equipment was provided for the full width of the LA. The long adder was segmented into partial adders of 64 bits. In Section 1.5.1 67, and in Section 1.5.2 10, such units were used. During an addition of a summand, however, in Section 1.5.1 only 4, and in Section 1.5.2 only 3, of these units are activated. This raises the question whether adder equipment is really needed for the full width of the LA and whether the accumulation can be done with only 4 or 3 adders in accordance with Solution B of Section 1.2.2. There the LA is kept as local memory on the arithmetic unit.

1.5 Scalar Product Units for Top-Performance Computers

In this section we develop such a solution for the double precision data format. An in-principle solution using a short adder and local memory on the arithmetic unit was discussed in Section 1.3.2. There the data a_i and b_i to perform a product $a_i \times b_i$ are read into the SPU successively in two portions of 64 bits. This leaves 4 machine *cycles* to perform the accumulation in the pipeline.

Now we assume that the two data a_i and b_i for a product $a_i \times b_i$ are read into the SPU simultaneously in one portion of 128 bits. Again we call the time that is needed for this a *cycle*. In accordance with the solution shown in Fig. 1.11 and Section 1.5.1 we assume again that the multiplication and the shift also can be done in one such read *cycle*. In a balanced pipeline, then, the circuit for the accumulation must be able to read and process one summand in each (read) *cycle* also. The circuit in Fig. 1.13 displays a solution. Closely following the summing matrix in Fig. 1.11 we assume there that the local memory LA is organized in 17 rows of four 64 bit words.

In each *cycle* the multiplier supplies a product (summand) to be added in the accumulation unit. Every such summand carries an exponent which in our example consists of 12 bits. The 8 lower (least significant) bits of the exponent determine the shift width. The row selection of the LA is obtained by the 4 most significant bits of the exponent. This roughly corresponds to the selection of the adding position in two steps by the process described in the context of Fig. 1.2. The shift width and the row selection for the addition of the product to the LA are known as soon as the exponent of the product has been computed. Since the exponents of a_i and b_i consist of 11 bits only, the result of their addition is available very quickly. So while the multiplication of the mantissa is still being executed the shifter can already be switched and the addresses for the LA words for the accumulation of the product $a_i \times b_i$ can be selected.

After being shifted the summand reaches the accumulation unit. It is read into the input register IR of this unit. The shifted summand now consists of an exponent e, a sign s, and a mantissa m. The mantissa touches three consecutive words of the LA, while the exponent is reduced to the four most significant bits of the original exponent of the product.

Now the addition of the summand is performed in the accumulation unit by the following three steps:

1. The local memory is addressed by the exponent e. The contents of the addressed part of the LA including the word which resolves the carry are transferred to the register before summation RBS. This transfer moves four words of 64 bits. The summand is also transferred from IR to the corresponding section of RBS. In Fig. 1.13 this part of the RBS is denoted by e', s' and m' respectively.
2. In the next *cycle* the addition or subtraction is executed in the add/subtract unit according to the sign. The result is transferred to the register after summation RAS. The adder/subtracter consists of 4 parallel adders

42   1. Fast and Accurate Vector Operations

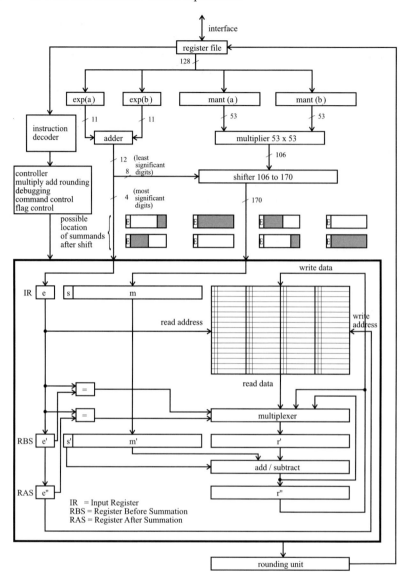

Fig. 1.13. Block diagram of a SPU with short adder and local store for a 64 bit data word and 128 bit data bus

1.5 Scalar Product Units for Top-Performance Computers 43

of 64 bits which are working in carry select mode. The summand touches three of these adders. Each one of these three adders can produce a carry. The carries between two of these adjacent adders are absorbed by the carry select addition. The fourth word is the carry word. It is selected by the flag mechanism. During the addition step a 1 is added to or subtracted from this word in carry select mode. If the addition produces a carry the incremented/decremented word will be selected. If the addition does not produce a carry this word remains unchanged. Simultaneously with the incrementation/decrementation of the carry word a second set of flags is set up which is copied into the flag word in the case that a carry is generated. In Fig. 1.13 two possible locations of the summand after the shift are indicated. The carry word is always the most significant word. An incrementation/decrementation of this word never produces a carry. Thus the adder/subtracter in Fig. 1.13 simply can be built as a parallel carry select adder.

3. In the next *cycle* the computed sum is written back into the same four memory cells of the LA to which the addition has been executed. Thus only one address decoding is necessary for the read and write step. A different bus called *write data* in Fig. 1.13 is used for this purpose.

In summary the addition consists of the typical three steps: 1. read the summand, 2. perform the addition, and 3. write the sum back into the (local) memory. Since a summand is delivered from the multiplier in each *cycle*, all three phases must be active simultaneously, i. e. the addition itself must be performed in a pipeline. This means that it must be possible to read from the memory and to write into the memory in each *cycle* simultaneously. So two different data paths have to be provided. This, however, is usual for register memory.

The pipeline for the addition consists of three steps. Pipeline conflicts are quite possible. A pipeline conflict occurs if an incoming summand needs to be added to a partner from the LA which is still being computed and not yet available in the local memory. These situations can be detected by comparing the exponents e, e' and e'' of three successively incoming summands. In principle all pipeline conflicts can be solved by the hardware. Here we discuss the solution of two pipeline conflicts which with high probability are the most frequent occurrences.

One conflict situation occurs if two consecutive products carry the same exponent e. In this case the two summands touch the same three words of the LA. Then the second summand is unable to read its partner for the addition from the local memory because it is not yet available. This situation is checked by the hardware where the exponents e and e' of two consecutive summands are compared. If they are identical, the multiplexer blocks off the process of reading from the local memory. Instead the sum which is just being computed is directly written back into the register before summation RBS via the multiplexer so that the second summand can immediately be added without memory involvement.

Another possibility of a pipeline conflict occurs if from three successively incoming summands the first one and the third one carry the same exponent. Since the pipeline consists of three steps, the partner for the addition of the third one then is not yet in the local memory but still in the register after summation RAS. This situation is checked by the hardware also, see Fig. 1.13. There the two exponents e and e'' of the two summands are compared. In case of coincidence the multiplier again suppresses the reading from the local memory. Instead now, the sum of the former addition, the result of which is still in RAS, is directly written back into the register RBS before summation via the multiplexer. So also this pipeline conflict can be solved by the hardware without memory involvement.

The case $e = e' = e''$ is also possible. It would cause a reading conflict in the multiplexer. The situation can be avoided by writing a dummy exponent into e'' or by reading from the add/subtract unit with higher priority.

The product that arrives at the accumulation unit touches three consecutive words of the LA. A more significant fourth word absorbs the possible carry. The solution for the two pipeline conflicts just described works well, if this fourth word is the next more significant word. A carry is not absorbed by the fourth word if all its bits are one, or are all zero. The probability that this is the case is $1 : 2^{64} < 10^{-18}$. In the vast majority of instances this will not be the case.

If it is the case the word which absorbs the carry is selected by the flag mechanism and read into the most significant word of the RBS. The addition step then again works well including the carry resolution. But difficulties occur in both cases of a pipeline conflict. Fig. 1.14 displays a certain part of the LA. The three words to which the addition is executed are denoted by 1, 2 and 3. The next more significant word is denoted by 4 and the word which absorbs the carry by 5.

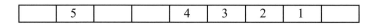

Fig. 1.14. Carry propagation in case of a pipeline conflict

In case of a pipeline conflict with $e = e'$ or $e = e''$ the following addition again touches the words 1, 2 and 3. Now the carry is absorbed either by word 4 or by word 5. Word 4 absorbs the carry if an addition is followed by an addition or a subtraction followed by a subtraction. Word 5 absorbs the carry if an addition is followed by a subtraction or vice versa. So the hardware has to take care that either word 4 or 5 is read into the most significant word of RBS depending on the operation which follows. The case that word 5 is the carry word again needs no particular care. Word 5 is already in the most significant position of the RBS. It is simply treated the same way as the words 1, 2 and 3. In the other case word 4 has to be read from the LA into RBS, simultaneously with the words 1, 2 and 3 from the add/subtract unit or from RAS into RBS. In this case word 5 is written into the local memory via the normal write path.

So far certain solutions for the possible pipeline conflicts $e = e'$ and $e = e''$ have been discussed. These are the most frequent but not the only conflicts that may occur. Similar difficulties appear if two or three successive incoming summands overlap only partially. In this case the exponents e and e' and/or e'' differ by 1 or 2 so that also these situations can be detected by comparison of the exponents. Another pipeline conflict appears if one of the two following summands overlaps with a carry word. In these cases summands have to be built up in parts from the adder/subtracter or RAS and the LA. Thus hardware solutions for these situations are more complicated and costly. We leave a detailed study of these situations to the reader/designer and offer the following alternative: The accumulation pipeline consists of three steps only. Instead of investing in a lot of hardware logic for rare situations of a pipeline conflict it may be simpler and less expensive to stall the pipeline and delay the accumulation by one or two *cycles* as needed. It should be mentioned that other details as for instance the width of the adder that is used also can heavily change the design aspects. A 128 instead of a 64 bit adder width which was assumed here could simplify several details.

It was already mentioned that the probability for the carry to run further than the fourth word is less than 10^{-18}. A particular situation where this happens occurs if the sum changes its sign from a positive to a negative value or vice versa. This can happen frequently. To avoid a complicated carry handling procedure in this case a small carry counter of perhaps three bits could be appended to each 64 bit word of the LA. If these counters are not zero at the end of the accumulation their contents have to be added to the LA. For further details see [66], [45].

As was pointed out in connection with the unit discussed in Section 1.3.2, the addition of the summand actually can be carried out over 170 bits only. Thus the shifter that is shown in Fig. 1.13 can be reduced to a 106 to 170 bits shifter and the data path from the shifter to the input register IR as well as the one to RBS also need to be 170 bits wide only. If this possible hardware reduction is applied, the summand has to be expanded to the full 256 bits when it is transferred to the adder/subtracter.

1.5.4 Short Adder with Local Memory on the Arithmetic Unit for 32 Bit Data Word (Solution B)

Now we consider again a 32 bit data word. We assume that two of these are read simultaneously into the SPU in one read *cycle*. The LA is kept as local memory in the SPU. We assume that the addition of a summand, which now is a 48 bit product, can be done by three adders of 64 bits including the carry resolution. Multiplication of the mantissas and addition of the exponents are done in full accordance with the upper part of the circuits shown in Fig. 1.12. The shift is executed similarly to the one in Fig. 1.12. We shall comment on it later. The appropriately shifted product then reaches the accumulation unit. A block diagram of this unit is shown in Fig. 1.15.

We assume that the multiplication and the shift can be performed in one read *cycle*. Then, a shifted product reaches the input register IR of the accumulation unit in each (read) *cycle*. The accumulation unit must add and process one summand in each such *cycle*. The addition itself is performed by the following three steps, see Fig. 1.15.

1. The product which is already in IR touches at most two successive 64 bit words of the LA. These words are addressed by the exponent e of the product. The contents of these two words of the LA and the word which absorbs the carry are transferred from the LA to the register part r' of RBS. This transfer moves three 64 bit words. The summand in IR is also transferred to the corresponding section of RBS. This part is denoted by e', s' and m' in Fig. 1.15.
2. In the next step the addition or subtraction is executed in the add/subtract unit according to the sign. The result is transferred to the register RAS. The adder/subtracter consists of three 64 bit adders which are working in carry select mode. So the carries between the lower two of these adders are absorbed by the carry select addition. The carry word is the most significant one. An incrementation/decrementation of this word never produces a carry. Thus the adder/subtracter in Fig. 1.15 can be built simply as a parallel adder.
3. In the next *cycle* the computed sum is written back into the same three memory cells of the LA to which the addition has been executed. The *write* bus is used for this purpose. Thus only one address decoding is necessary for the read and write step.

Since a summand is delivered from the multiplier in each *cycle*, all three of these phases must be active simultaneously, i. e. the addition must be performed in a pipeline. This means, in particular, that it must be possible to read from the LA and to write into the LA simultaneously in each *cycle*. Therefore, two different data paths have to be provided, as shown in Fig. 1.15.

The pipeline for the addition consists of three steps. Pipeline conflicts again are quite possible. A pipeline conflict occurs if an incoming summand needs to be added to a partner from the LA which is still being computed and not yet available in the local memory. These situations can be detected by comparing the exponents e, e' and e'' of three successively incoming summands. In principle all pipeline conflicts can be solved by the hardware. We discuss here the solution of two pipeline conflicts which with high probability are the most frequent occurrences.

1.5 Scalar Product Units for Top-Performance Computers

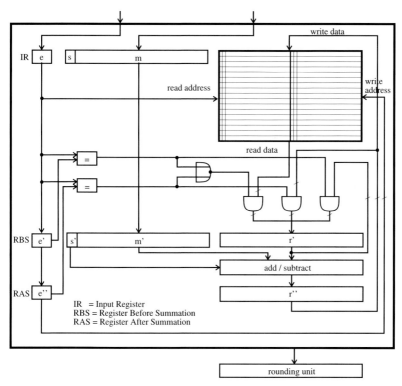

Fig. 1.15. Block diagram for a SPU with short adder and local store for a 32 bit data word and 64 bit data bus

One conflict situation occurs if two consecutive products carry the same exponent e. In this case the two summands touch the same two words of the LA. Then the second summand is unable to read its partner for the addition from the LA because it is not yet available. This situation is checked by the hardware where the exponent e and e' of two consecutive summands are compared. In case of coincidence the process of reading from the LA is blocked off. Instead the sum which is just being computed is directly written back into the register RBS so that the second summand can immediately be added without memory involvement.

Another possibility of a pipeline conflict occurs if from three successive incoming summands the first one and the third one carry the same exponent. Since the pipeline consists of three phases the partner for the addition of the third one then is not yet in the LA but still in the register RAS. This situation is checked by the hardware as well, see Fig. 1.15. There the two exponents e and e'' are compared. In case of coincidence again the process of reading from the LA is blocked off. Instead now, the result of the former

48   1. Fast and Accurate Vector Operations

addition, which is still in RAS, is directly written back into RBS. Then the addition can be executed without LA involvement.

The case $e = e' = e''$ is also possible. It would cause a conflict in the selection unit which in Fig. 1.15 is shown directly beneath of the LA. The situation can be avoided by writing a dummy exponent into e'' or by reading from the add/subtract unit with higher priority. This solution is not shown in Fig. 1.15.

The product that arrives at the accumulation unit touches two consecutive words of the LA. A more significant third word absorbs the possible carry. The solution for the two pipeline conflicts work well, if this third word is the next more significant word of the LA. The probability that this is not the case is less than 10^{-18}. In the vast majority of instances this will be the case.

If it is not the case the word which absorbs the carry is selected by the flag mechanism and read into the most significant word of the RBS. The addition step then works well again including the carry resolution. But difficulties can occur in both cases of a pipeline conflict. Fig. 1.16 shows a certain part of the LA. The two words to which the addition is executed are denoted by 1 and 2. The next more significant word is denoted by 3 and the word which absorbs the carry by 4.

Fig. 1.16. Carry propagation in case of a pipeline conflict

In case of a pipeline conflict with $e = e'$ or $e = e''$ the following addition again touches the words 1 and 2. Now the carry is absorbed either by the word 3 or by the word 4. Word 3 absorbs the carry if an addition is followed by an addition or a subtraction is followed by a subtraction. Word 4 absorbs the carry if an addition is followed by a subtraction or vice versa. So the hardware has to take care that either word 3 or word 4 is read into the most significant word of RBS depending on the operation which follows. The case that the word 4 is the carry word again needs no particular care. Word 4 is already in the most significant position of the RBS. It is simply treated the same way as the words 1 and 2. In the other case word 3 has to be read from the LA into RBS simultaneously with the words 1 and 2 from the add/subtract unit or from RAS into RBS. In this case word 4 is written into the local memory via the normal write path.

So far solutions for the two pipeline conflicts $e = e'$ and $e = e''$ have been discussed. These are not the only conflicts that may occur. Similar difficulties appear if two or three successively incoming summands overlap only partially. In this case the exponents e and e' and/or e'' differ by 1 so that these situations can be detected by comparison of the exponents also. Another pipeline conflict appears if one of the following two summands overlaps with a carry word. In these cases summands have to be built up in parts from the adder/subtracter or RAS and the LA. Thus hardware solutions for these situations are more complicated and costly. We leave a detailed study of these situations to the reader/designer and offer the following alternative. The accumulation pipeline consists of three steps only. Instead of investing in a lot of hardware logic for very rare situations of a pipeline conflict it may be simpler and less expensive to stall the pipeline and delay the accumulation by one or two *cycles* as needed.

The product consists of 48 bits. So the summand never touches the 16 least significant bits of word 1. The most significant third 64 bit word of the adder is supposed to absorb the carry. It can be built as an incrementer/decrementer by halfadders. Thus, in comparison with Fig. 1.12, the shifter can be reduced to a 48 to 112 bit shifter and the data path from the shifter to the input register IR as well as the one to RBS also needs to be 112 bits wide only. If this possibility is chosen, the summand has to be expanded to the full 192 bits when it is read into the adder/subtracter.

The circuits that have been discussed so far are based on the assumption that the LA is organized in words of 64 bits and that the partial adder that is used is also 64 bits wide. It should be mentioned that these assumptions, although realistic, are nevertheless somewhat arbitrary and that other choices are quite possible and may lead to simpler or better solutions. The LA could as well be organized in words of 128 or only 32 bits. The width of the partial adder could also be 128 or 32 bits. All these possibilities allow interesting solutions for the different cases that have been discussed in this paper. We leave it to the reader to play with these combinations and select the one which fits best to a given hardware environment. With increasing word size the probability for a pipeline conflict which has not been discussed so far decreases.

1.6 Hardware Accumulation Window

So far it has been assumed in this paper that the SPU is incorporated as an integral part of the arithmetic unit of the processor. Now we discuss the question of what can be done if not enough register space for the LA is available on the processor.

The final result of a scalar product computation is assumed to be a floating-point number with an exponent in the range $e1 \leq e \leq e2$. If this is not the case, the problem has to be scaled. During the computation of the scalar product, however, summands with an exponent outside of this range may occur. The remaining computation then has to cancel all the digits outside of the range $e1 \leq e \leq e2$. So in a normal scalar product computation, the register space outside this range will be used less frequently. It was already mentioned earlier in this paper that the conclusion should not be drawn from this consideration that the register size can be restricted to the

50   1. Fast and Accurate Vector Operations

single exponent range in order to save some silicon area. This would require the installment of complicated exception handling routines in software or in hardware. The latter may finally require as much silicon. A software solution certainly is much slower. The hardware requirement for the LA in case of standard arithmetics is modest and the necessary register space really should be invested.

However, the memory space for the LA on the arithmetic unit grows with the exponent range of the data format. If this range is extremely large, as for instance in case of an extended precision floating-point format, then only an inner part of the LA can be supported by hardware. We call this part of the LA a Hardware Accumulation Window (HAW). See Fig. 1.17. The outer parts of this window must then be handled in software. They are probably needed less often.

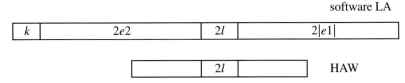

Fig. 1.17. Hardware Accumulation Window (HAW)

There are still other reasons that suppose the development of techniques for the computation of the accurate scalar product using a HAW. Many conventional computers on the market do not provide enough register space to represent the full LA on the CPU. Then a HAW is one choice which allows a fast and correct computation of the scalar product in many cases.

Another possibility is to place the LA in the user memory, i. e. in the data cache. In this case only the start address of the LA and the flag bits are put into (fixed) registers of the general purpose register set of the computer. This solution has the advantage that only a few registers are needed and that a longer accumulator window or even the full LA can be provided. This reduces the need to handle exceptions. The disadvantage of this solution is that for each accumulation step, four memory words must be read and written in addition to the two operand loads. So the scalar product computation speed is limited by the data cache to processor transfer bandwidth and speed. If the full long accumulator is provided this is a very natural solution. It has been realized on several IBM, SIEMENS and HITACHI computers of the /370 architecture in the 1980s [109, 110, 112, 119].

A faster solution certainly is obtained for many applications with a HAW in the general purpose register set of the processor. Here only a part of the LA is present in hardware. Overflows and underflows of this window have to be handled by software. A full LA for the data format double precision of the IEEE-arithmetic standard 754 requires 4288 bits or 67 words of 64 bits. We assume here that only 10 of these words are located in the general purpose register set.

Such a window covers the full LA that is needed for a scalar product computation in case of the data format single precision of the IEEE-arithmetic standard 754. It also allows a correct computation of scalar products in the case of the long data format of the /370 architecture as long as no under- or overflows occur. In this case $64 + 28 + 63 = 155$ hexadecimal digits or 620 bits are required. With a HAW of 640 bits all scalar products that do not cause an under- or overflow could have been correctly computed on these machines. This architecture was successfully used and even dominated the market for more than 20 years. This example shows that even if a HAW of only 640 bits is available, the vast majority of scalar products will execute on fast hardware.

Of course, even if only a HAW is available, all scalar products should be computed correctly. Any operation that over- or underflows the HAW must be completed in software. This requires a complete software implementation of the LA, i. e. a variable of type dotprecision. All additions that do not fit into the HAW must be executed in software into this dotprecision variable.

There are three situations where the HAW can not correctly accumulate the product:

- the exponent of the product is so high that the product does not (completely) fit into the HAW. Then the product is added in software to the dotprecision variable.
- the exponent of the product is so low that the product does not (completely) fit into the HAW. Then the product is added in software to the dotprecision variable.
- the product fits into the HAW, but its accumulation causes a carry to be propagated outside the range of the HAW. In this case the product is added into the HAW. The carry must be added in software to the dotprecision variable.

If at the end of the accumulation the contents of the software accumulator are non zero, the contents of the HAW must be added to the software accumulator to obtain the correct value of the scalar product. Then a rounding can be performed if required. If at the end of the accumulation the contents of the software accumulator are zero, the HAW contains the correct value of the scalar product and a rounded value can be obtained from it.

Thus, in general, a software controlled full LA supplements a HAW. The software routines must be able to perform the following functions:

- clear the software LA. This routine must be called during the initialization of the HAW. Ideally, this routine only sets a flag. The actual clearing is only done if the software LA is needed.
- add or subtract a product to/from the software LA.

- add or subtract a carry or borrow to/from the software LA at the appropriate digit position.
- add the HAW to the software LA. This is required to produce the final result when both the HAW and the software LA were used. Then a rounding can be performed.
- round the software LA to a floating-point number.

With this software support scalar products can be computed correctly using a HAW at the cost of a substantial software overhead and a considerable time penalty for products that fall outside the range of the HAW.

An alternative to the HAW-software environment just described is to discard the products that underflow the HAW. A counter variable is used to count the number of discarded products. If a number of products were discarded, the last bits of the HAW must be considered invalid. A valid rounded result can be generated by hardware if these bits are not needed. If this procedure fails to produce a useful answer the whole accumulation is repeated in software using a full LA.

A 640 bit HAW seems to be the shortest satisfactory hardware window. If this much register space is not available, a software implementation probably is the best solution.

If a shorter HAW must be implemented, then it should be a movable window. This can be represented by an exponent register associated with the hardware window. At the beginning of an accumulation, the exponent register is set so that the window covers the least significant portion of the LA. Whenever a product would cause the window to overflow, its exponent tag is adjusted, i. e. the window moves to the left, so that the product fits into the window. Products that would cause an underflow are counted and otherwise ignored. The rounding instruction checks whether enough significant digits are left to produce a correctly rounded result or whether too much cancellation did occur. In the latter case it is up to the user to accept the inexact result or to repeat the whole accumulation in software using a full LA.

Using this technique a HAW as short as 256 bits could be used to perform rounded scalar product computation and quadruple precision arithmetic. However, it would not be possible to perform many other nice and useful applications of the optimal scalar product with this type of scalar product hardware as for instance a *long real arithmetic*.

The software overhead caused by the reduction of the full width of the LA to a HAW represents a trade off between hardware expenditure and runtime.

With the accurate scalar product operators for multiple precision arithmetic including multiple precision interval arithmetic can easily be provided. This enables the user to use higher precision operations in numerically critical parts of a computation. Experience shows that if one runs out of precision in a certain problem class one often runs out of *double* or *extended* precision very soon as well. It is preferable and simpler, therefore, to provide the principles for enlarging the precision than simply providing any fixed higher precision.

To allow fast execution of a number of multiple precision arithmetics the HAW should not be too small.

1.7 Theoretical Foundation of Advanced Computer Arithmetic and Shortcomings of Existing Processors and Standards

Arithmetic is the basis of mathematics. Advanced computer arithmetic expands the arithmetic and mathematical capability of the digital computer in the most natural way. Instead of reducing all calculations to the four elementary operations for floating-point numbers, advanced computer arithmetic provides twelve fundamental data types or mathematical spaces with operations of highest accuracy in a computing environment.

Besides the real numbers, the complex numbers form the basis of analysis. For computations with guarantees one needs the intervals over the real and complex numbers as well. The intervals bring the continuum onto the computer. An interval between two floating-point bounds represents the continuous set of real numbers between these two bounds.

The twelve fundamental data types or mathematical spaces consist of the four basic data types real, complex, interval and complex interval as well as the vectors and matrices over these types. See Fig. 1.18 and Fig. 1.19. Arithmetic operations in the computer representable subsets of these spaces are defined by a general mapping principle which is called a semimorphism. These arithmetic operations are distinctly different from the customary ones in these spaces which are based on elementary floating-point arithmetic.

If M is any one of these twelve data types (or mathematical spaces) and N is its computer representable subset, then for every arithmetic operation $\circ$ in M, a corresponding computer operation $\boxdot$ in N is defined by

(RG) $\quad a \boxdot b := \Box(a \circ b) \quad$ for all $a, b \in N$ and all operations $\circ$ in M,

where $\Box : M \to N$ is a mapping from M onto N which is called a rounding if it has the following properties:

(R1) $\quad \Box a = a \quad\quad$ for all $a \in N \quad\quad$ (projection).
(R2) $\quad a \leq b \Rightarrow \Box a \leq \Box b \quad\quad$ for $a, b \in M \quad\quad$ (monotonicity).

The concept of semimorphism requires additionally that the rounding is antisymmetric, i. e. that it has the property

(R3) $\quad \Box(-a) = -\Box(a) \quad\quad$ for all $a \in M \quad\quad$ (antisymmetry).

For the interval spaces among the twelve basic data types — the intervals over the real and complex numbers as well as the intervals over the real and complex vectors and matrices — the order relation in (R2) is the subset relation $\subseteq$. A rounding from any interval set M onto its computer representable subset N is defined by properties (R1), (R2) (with $\leq$ replaced by $\subseteq$), plus the additional property

(R4) $a \subseteq \square a$ for all $a \in M$ (inclusion).

These interval roundings are also antisymmetric, that is, they satisfy property (R3) [60, 62].

Additional important roundings from the real numbers onto the floating-point numbers are the montone downwardly and upwardly directed roundings with the property

(R4) $\nabla a \leq a$ resp. $a \leq \triangle a$ for all $a \in M$ (directed).

These directed roundings are uniquely defined by (R1), (R2) and (R4), see [60, 62]. Arithmetic operations are also defined by (RG) with the roundings ∇ and $\triangle$.

With the five rules (RG) and (R1, 2, 3, 4), a large number of arithmetic operations is defined in the computer representable subsets of the twelve fundamental data types or mathematical spaces. (RG) means that every computer operation should be performed in such a way that it produces the same result as if the mathematically correct operation were first performed in the basic space M and the exact result then rounded into the computer representable subset N. In contrast to the traditional approximation of the arithmetic operations in the product spaces by floating-point arithmetic, all operations with the properties (RG), (R1) and (R2) are optimal in the sense that there is no better computer representable approximation to the true result (with respect to the prescribed rounding). In other words, between the correct and the computed result of an operation there is no other element of the corresponding computer representable subset. This can easily be seen: Let $a, b \in N$, and $\alpha \in N$ the greatest lower and $\beta \in N$ the least upper bound of the correct result $a \circ b$ in M, i. e.

$$\alpha \leq a \circ b \leq \beta,$$

then

$$\underset{(R1)}{\square \alpha} = \underset{(R2)}{\alpha} \leq \underset{(RG)}{\square(a \circ b)} = \underset{(R2)}{a \boxdot b} \leq \underset{(R1)}{\square \beta} = \beta. \qquad (1.1)$$

Thus, all semimorphic computer operations are of 1 ulp (<u>u</u>nit in the <u>l</u>ast <u>p</u>lace) accuracy. 1/2 ulp accuracy is achieved in the case of rounding to nearest. In the product spaces the order relation is defined componentwise. So in the product spaces property (1.1) holds for every component.

Figure 1.18 shows a table of the twelve basic arithmetic data types and corresponding operators as they are provided by the programming language PASCAL-XSC [46, 47, 49, 67, 68, 108]. All data types and operators are predefined available in the language. The operations can be called by the operator symbols shown in the table. An arithmetic operator followed by a less or greater symbol denotes an operation with rounding downwards or upwards, respectively. The operator +* takes the interval hull of two elements, ** means intersection. Also all outer operations that occur in Fig. 1.18 (scalar times vector, matrix times vector, etc.) are defined by the five properties

1.7 Theoretical Foundation of Advanced Computer Arithmetic

(RG), (R1, 2, 3, 4), whatever applies. A count of all inner and outer predefined operations in the figure leads to a number of about 600 arithmetic operations.

left operand \ right op.	integer real complex	interval cinterval	rvector cvector	ivector civector	rmatrix cmatrix	imatrix cimatrix
monadic	+, −	+, −	+, −	+, −	+, −	+, −
integer real complex	+, +<, +> −, −<, −> *, *<, *> /, /<, /> +*	+, −, *, / +*	*, *<, *>	*	*, *<, *>	*
interval cinterval	+, −, *, / +*	+, −, *, / +*, **	*	*	*	*
rvector cvector	*, *<, *> /, /<, />	*, /	+, +<, +> −, −<, −> *, *<, *> +*	+, −, * +*		
ivector civector	*, /	*, /	+, −, * +*	+, −, * +*, **		
rmatrix cmatrix	*, *<, *> /, /<, />	*, /	*, *<, *>	*	+, +<, +> −, −<, −> *, *<, *> +*	+, −, * +*
imatrix cimatrix	*, /	*, /	*	*	+, −, * +*	+, −, * +*, **

Fig. 1.18. Predefined arithmetic data types and operators of PASCAL-XSC.

Figure 1.19 lists the same data types in their usual mathematical notation. There $I\!R$ denotes the real and C the complex numbers. A heading letter V, M and I denotes vectors, matrices and intervals, respectively. R stands for the set of floating-point numbers and D for any set of higher precision floating-point numbers. If M is any set, $I\!P M$ denotes the power set, which is the set of all subsets of M. For any operation $\circ$ in M a corresponding operation $\circ$ in $I\!P M$ is defined by $A \circ B := \{a \circ b \mid a \in A \wedge b \in B\}$ for all $A, B \in I\!P M$.

For each set-subset pair in Fig. 1.19, arithmetic in the subset is defined by semimorphism. These operations are different in general from those which are performed in the product spaces if only elementary floating-point arithmetic is furnished on the computer. Semimorphism defines operations in a subset N of a set M directly by making use of the operations in M. It makes a direct link between an operation in M and its approximation in the subset N. For instance, the operations in MCR (see Fig. 1.19) are directly defined by the operations in MC, and not in a roundabout way via C, $I\!R$, R, CR, and MCR as it would have to be done by using elementary floating-point arithmetic only.

$$
\begin{array}{cccc}
 & & \mathbb{R} \supset & D \supset & R \\
 & & V\mathbb{R} \supset & VD \supset & VR \\
 & & M\mathbb{R} \supset & MD \supset & MR \\
\mathbb{P}\mathbb{R} \supset & I\mathbb{R} \supset & ID \supset & IR \\
\mathbb{P}V\mathbb{R} \supset & IV\mathbb{R} \supset & IVD \supset & IVR \\
\mathbb{P}M\mathbb{R} \supset & IM\mathbb{R} \supset & IMD \supset & IMR \\
 & & \mathbb{C} \supset & CD \supset & CR \\
 & & V\mathbb{C} \supset & VCD \supset & VCR \\
 & & M\mathbb{C} \supset & MCD \supset & MCR \\
\mathbb{P}\mathbb{C} \supset & I\mathbb{C} \supset & ICD \supset & ICR \\
\mathbb{P}V\mathbb{C} \supset & IV\mathbb{C} \supset & IVCD \supset & IVCR \\
\mathbb{P}M\mathbb{C} \supset & IM\mathbb{C} \supset & IMCD \supset & IMCR \\
\end{array}
$$

Fig. 1.19. Table of the spaces occurring in numerical computations.

The properties of a semimorphism can be derived as necessary conditions for an homomorphism between ordered algebraic structures [60, 62]. It is easy to see that repetition of semimorphism is again a semimorphism. A careful analysis of the requirements of semimorphism is given in [60, 62]. The resulting algebraic and order structure are studied there under the mapping properties (RG) and (R1, 2, 3, 4). Many properties of both the order structure and the algebraic structure are invariant under a semimorphism. Because of (R2) with respect to $\leq$ or $\subseteq$ the order structure is not changed if we move from a set into a subset in any row of Fig. 1.19, while the algebraic structure is considerably weakened. The concept of semimorphism and its explicit five rules (RG), (R1, 2, 3, 4) are used as an axiomatic definition of computer arithmetic in the XSC-languages [14, 41, 46–49, 56, 67, 68, 106–108, 112].

In the theory of computer arithmetic it is ultimately shown, that all arithmetic operations of the twelve fundamental numerical data types of Fig. 1.18 or spaces of Fig. 1.19 can be provided in a higher programming language by a modular technique, if on a low level, preferably in hardware, 15 fundamental operations are available: the five operations $+, -, \times, /, \cdot$, each one with the three roundings $\square, \triangledown, \triangle$. Here · means the scalar product of two vectors, $\square$ is a monotone, antisymmetric rounding, e. g. rounding to nearest, and $\triangledown$ and $\triangle$ are the monotone downwardly and upwardly directed roundings from the real numbers into the floating-point numbers. All 15 operations $\square, \triangledown, \triangle$, with $\circ \in \{+, -, \times, /, \cdot\}$, Fig. 1.20, are defined by (RG). In case of the scalar product, a and b are vectors $a = (a_i)$, $b = (b_i)$ with any finite number of components.

1.7 Theoretical Foundation of Advanced Computer Arithmetic

$$\boxplus, \quad \boxminus, \quad \boxtimes, \quad \boxslash, \quad \boxdot, \qquad a \boxdot b = \boxdot \sum_{i=1}^{n} a_i \times b_i,$$

$$\triangledown, \quad \triangledown, \quad \triangledown, \quad \triangledown, \quad \triangledown, \qquad a \triangledown b = \triangledown \sum_{i=1}^{n} a_i \times b_i,$$

$$\triangle, \quad \triangle, \quad \triangle, \quad \triangle, \quad \triangle, \qquad a \triangle b = \triangle \sum_{i=1}^{n} a_i \times b_i.$$

Fig. 1.20. The fifteen fundamental operations for advanced computer arithmetic.

The IEEE arithmetic standards 754 and 854 offer 12 of these operations: $\boxdot, \triangledown, \triangle$, with $\circ \in \{+, -, \times, /\}$. These standards also prescribe specific data formats. A general theory of computer arithmetic is not bound to these data formats. By adding just three more operations, the optimal scalar products $\boxdot, \triangledown, \triangle$, all operations in the usual product spaces of numerical mathematics can be performed with 1 or 1/2 ulp accuracy in each component.

Remark 1: With this information it seems to be relatively easy to provide advanced computer arithmetic on processors which offer the IEEE arithmetic standard 754. The standard seems to be a step in the right direction. All that is additionally needed are the three optimal scalar products $\boxdot$, $\triangledown$ and $\triangle$. If they are not supported by the computer hardware they could be simulated. One possibility to simulate these operations certainly would be to place the LA into the user memory, i. e. in the data cache. This possibility was discussed in Section 1.6.

However, a closer look into the subject reveals severe difficulties and disadvantages which result in unnecessary performance penalties. So that at a place where an increase in speed is to be expected, a severe loss of speed results instead.

A first severe drawback comes from the fact that processors that provide IEEE arithmetic separate the rounding from the operation. First the rounding mode has to be set. Then an arithmetic operation can be performed. In a conventional floating-point computation this does not cause any difficulties. The rounding mode is set only once. Then a large number of arithmetic operations is performed with this rounding mode. However, when interval arithmetic is performed, the rounding mode has to be switched very frequently. In the computer the lower bound of the result of every interval operation has to be rounded downwards and the upper bound rounded upwards. Thus the rounding mode has to be set for every arithmetic operation. If setting the rounding mode and the arithmetic operation are equally fast, this slows down interval arithmetic unnecessarily by a factor of two in comparison to a conventional floating-point arithmetic. On the Pentium processor setting the rounding mode takes three cycles, the following operation only one!! Thus an interval operation is 8 times slower than the corresponding floating-point operation. On workstations the situation is even worse in general. The rounding should be part of the arithmetic operation as required by the postulate

(RG) of the axiomatic definition of (advanced) computer arithmetic. Every one of the rounded operations $\square, \triangledown, \triangle, \circ \in \{+, -, \times, /\}$ should be executed in a single cycle! The rounding must be an integral part of the operation.

A second severe drawback comes from the fact that all the commercial processors that perform IEEE-arithmetic in case of multiplication only deliver a rounded product to the outside world. Computation of an accurate scalar product requires products of the full double length. These products have to be simulated from outside on the processor. This slows down the multiplication by a factor of 10 in comparison to a rounded hardware multiplication. In a software simulation of the accurate scalar product the products of double length then have to be accumulated into the LA. This process is again slower by a factor of 5 in comparison to a (possibly wrong) hardware accumulation of products in floating-point arithmetic. Thus in summary a factor of at least 50 for the runtime is the trade-off for an accurate computation of the scalar product on existing processors. This is too much to be easily accepted by the user. Again at a place where an increase in speed by a factor of between two and four is to be expected if the scalar product is supported by hardware, a severe loss of speed is obtained by processors which have not been designed for accurate computation of the scalar product.

A third severe drawback is the fact that no reasonable interface to the programming languages is required by existing computer arithmetic standards. The majority of operations shown in Fig. 1.18 can be provided in a programming language which allows operator overloading. Operator overloading, however, is not enough to call the twelve operations $\square, \triangledown, \triangle, \circ \in \{+, -, \times, /\}$ which are provided by all IEEE-arithmetic processors in a higher programming language. A general operator concept is necessary for ease of programming (three real operations for $+, -, \times, /$). This solution has been chosen in PASCAL-XSC. In C-XSC which has been developed as a C++ class library, the 8 operators $\triangledown$ and $\triangle$, $\circ \in \{+, -, \times, /\}$ are hidden in the interval operations and not openly available. This is necessary because C++ does not allow three different operators for addition, subtraction, multiplication and division for the data type real.

Computer arithmetic is an integral part of all programming languages. The quality of the arithmetic operations should be an integral part of the definition of all programming languages. This can easily be done. All operations that are shown in Fig. 1.18 can be defined by the five simple rules (RG) and (R1, 2, 3, 4). In particular the eight operations $\triangledown$ and $\triangle$, with $\circ \in \{+, -, \times, /\}$ are defined by (RG), (R1), (R2) and (R4). All interval operations are defined by (RG), (R1), (R2), (R3) and (R4). All other operations that appear in Fig. 1.18 can be defined by (RG), (R1), (R2) and (R3) with the additional information whether rounding to nearest, towards infinity or towards zero is required. A precise definition of advanced computer arithmetic thus turns out to be short and simple.

1.7 Theoretical Foundation of Advanced Computer Arithmetic

IEEE-arithmetic has been developed as a standard for microprocessors in the early eighties at a time when the microprocessor was the 8086. Since that time the speed of microprocessors has been increased by several magnitudes. IEEE-arithmetic is now even provided and used by super computers, the speed of which is faster again by several magnitudes. All this is no longer in balance. With respect to arithmetic many manufacturers believe that realization of the IEEE-arithmetic standard is all that is necessary to do. In this way the existing standards prove to be a great hindrance to further progress. Advances in computer technology are now so profound that the arithmetic capability and repertoire of computers should be expanded to prepare the digital computer for the computations of the next century. The provision of Advanced Computer Arithmetic is the most natural way to do this.

Remark 2: A vector arithmetic coprocessor chip XPA 3233 for the PC has been developed in a CMOS 0.8 μm VLSI gate array technology at the author's Institute in 1993/94. VHDL and COMPASS design tools were used. For design details see [8, 43] and in particular [9]. The chip is connected with the PC via the PCI-bus. The PCI- and EMC-interface are integrated on chip. In its time the chip computed the accurate scalar product between two and four times faster than the PC an approximation in floating-point arithmetic. With increasing clock rate of the PC the PCI-bus turned out to be a severe bottle neck. To keep up with the increased speed the SPU must be integrated into the arithmetic logical unit of the processor and interconnected by an internal bus system.

The chip, see Fig. 1.21, realizes the SPU that has been discussed in Section 1.3.1, 207,000 transistors are needed. About 30% of the transistors and the silicon area are used for the local memory and the flag registers with the carry resolution logic. The remaining 70% of the silicon area is needed for the PCI/EMC-interface and the chip's own multiplier, shifter, adder and rounding unit. All these units would be superfluous if the SPU were integrated into the arithmetic unit of the processor. A multiplier, shifter, adder and rounding unit are already there. Everything just needs to be arranged a little differently. Thus finally the SPU requires fewer transistors and less silicon area than is needed for the exception handling of the IEEE-arithmetic standard. Logically the SPU is much more regular and simpler. With it a large number of exceptions that can occur in a conventional floating-point computation are avoided.

Testing of the coprocessor XPA 3233 was easy. XSC-languages had been available and used since 1980. There, an identical software simulation of the accurate scalar product had been implemented. Additionally a large number of problem solving routines had been developed and collected in the toolbox volumes [34, 35, 57]. All that had to be done was to change the PASCAL-XSC compiler a little to call the hardware chip instead of its software simulation. Surprisingly 40% of the chips on the first wafer were correct and, probably due to the high standard of the implementors and their familiarity with the theoretical background, with PASCAL-XSC and the toolbox routines no redesign was necessary. The chips produced identical results than the software simulation.

Modern computer technology can provide millions of transistors on a single chip. This allows solutions to put into the computer hardware which even an experienced computer user is totally unaware of. Due to the insufficient knowledge and familiarity with the technology, the design tools and implementation techniques, obvious and easy solutions are not demanded by mathematicians. The engineer on the other hand, who is familiar with these techniques, is not aware of the consequences for mathematics [37].

Remark 3: In addition to the numerical data types and operators displayed in Fig. 1.18, the XSC-languages provide an array type *staggered (staggered precision)* [89, 90] for multiple precision data. A variable of type *staggered* consists of an array of variables of the type of its components. Components of the *staggered* type can be of type *real* or of type *interval*. The value of a variable of type *staggered* is the sum of its components. Addition and subtraction of such multiple precision data can easily be performed in the LA. Multiplication of two variables of this type can be computed easily and fast by the accurate scalar product. Division is performed iteratively. The multiple precision data type *staggered* is controlled by a global variable called *stagprec*. If *stagprec* is 1, the *staggered* type is identical to its component type. If, for instance, *stagprec* is 4 each variable of this type consists of an array of four variables of its component type. Again its value is the sum of its components. The global variable *stagprec* can be increased or decreased at any place in a program. This enables the user to use higher precision data and operations in numerically critical parts of his computation. It helps to increase software reliability. The elementary functions for the type *staggered* are also available in the XSC-languages for the component types *real* and *interval* [22, 53]. In the case that *stagprec* is 2, a data type is encountered which occasionally is denoted as double-double or quadruple precision.

1.7 Theoretical Foundation of Advanced Computer Arithmetic 61

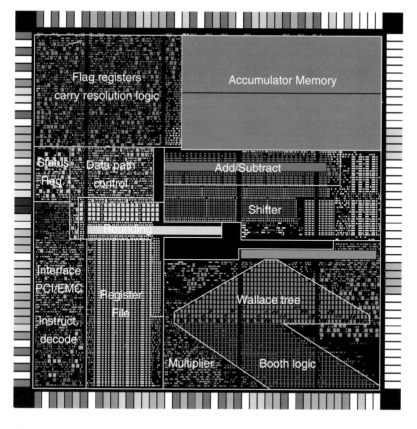

Fig. 1.21. Functional units, chip and board of the vector arithmetic coprocessor XPA 3233

Fig. 1.22. MADAS, model 20 BTZG, by H.W. Egli, Zürich, Switzerland
(**M**ultiplication, **A**utomatic **D**ivision, **A**ddition, **S**ubtraction)
0: Multiplication Register, **I**: Main or Product Register,
II: Counter or Dividend Register, **II**: Keybord or Entry Register,
IV: Accumulation Register.

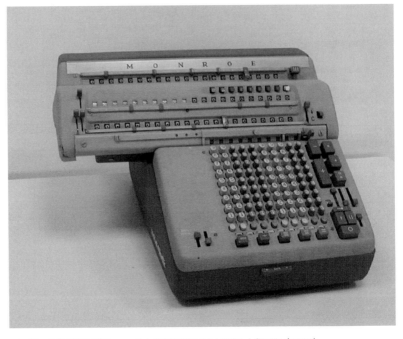

Fig. 1.23. MONROE, model MONROMATIC ASMD (1956),
by Monroe Calculating Machine Company, Inc., Orange, New Jersey, USA.
Addition, Subtraction, Multiplication, Division, Multiply and Accumulate.

Bibliography and Related Literature

1. Adams, E.; Kulisch, U.(eds.): **Scientific Computing with Automatic Result Verification.** I. Language and Programming Support for Verified Scientific Computation, II. Enclosure Methods and Algorithms with Automatic Result Verification, III. Applications in the Engineering Sciences. Academic Press, San Diego, 1993 (ISBN 0-12-044210-8).
2. Albrecht, R.; Kulisch, U. (Eds.): **Grundlagen der Computerarithmetik.** Computing Supplementum 1. Springer-Verlag, Wien / New York, 1977.
3. Albrecht, R.; Alefeld, G.; Stetter, H.J. (Eds.): **Validation Numerics – Theory and Applications.** Computing Supplementum 9, Springer-Verlag, Wien / New York, 1993.
4. Alefeld, G.; Herzberger, J.: *Einführung in die Intervallrechnung.* Bibliographisches Institut (Reihe Informatik, Nr. 12), Mannheim / Wien / Zürich, 1974 (ISBN 3-411-01466-0).
5. Alefeld, G.; Herzberger, J.: **An Introduction to Interval Computations.** Academic Press, New York, 1983 (ISBN 0-12-049820-0).
6. Apostolatos, N.; Kulisch, U.; Krawczyk, R.; Lortz, B.; Nickel, K.; Wippermann, H.-W.: *The Algorithmic Language Triplex-ALGOL 60.* Numerische Mathematik **11**, pp. 175-180, 1968.
7. Baumhof, Ch.: *Behavioural Description of A Scalar Product Unit.* Universität Karlsruhe, ESPRIT Project OMI/HORN, Deliverable Report D1.2/2, Dec. 1992.
8. Baumhof, Ch.: *A New VLSI Vector Arithmetic Coprocessor for the PC.* In [105, Vol. 12, pp. 210-215], 1995.
9. Baumhof, Ch.: *Ein Vektorarithmetik-Koprozessor in VLSI-Technik zur Unterstützung des Wissenschaftlichen Rechnens.* Dissertation, Universität Karlsruhe, 1996.
10. Baumhof, Ch.; Bohlender, G.: *A VLSI Vector Arithmetic Coprocessor for the PC.* Proceedings of WAI'96 in Recife/Brasil, RITA (Revista de Informática Teórica e Aplicada), Extra Edition, October 1996.
11. De Beauclair, W.: **Rechnen mit Maschinen.** Vieweg, Braunschweig, 1968.
12. Bierlox, N.: *Ein VHDL Koprozessor für das exakte Skalarprodukt.* Dissertation, Universität Karlsruhe, 2002.
13. Bleher, J. H.; Kulisch, U.; Metzger, M.; Rump, S. M.; Ullrich, Ch.; Walter, W.: *FORTRAN–SC: A Study of a FORTRAN Extension for Engineering/Scientific Computation with Access to ACRITH.* Computing **39**, pp. 93-110, Nov. 1987.
14. Blomquist, F.: **PASCAL-XSC, BCD-Version 1.0, Benutzerhandbuch für das dezimale Laufzeitsystem.** Universität Karlsruhe, Institut für Angewandte Mathematik, 1997.
15. Bohlender, G.: *Floating-Point Computation of Functions with Maximum Accuracy.* IEEE Transactions on Computers, Vol. C-26, no. 7, July 1977.

16. Bohlender, G.: *Genaue Berechnung mehrfacher Summen, Produkte und Wurzeln von Gleitkommazahlen und allgemeine Arithmetik in höheren Programmiersprachen.* Dissertation, Universität Karlsruhe, 1978.
17. Bohlender, G.; Grüner, K.; Kaucher, E.; Klatte, R.; Krämer, W.; Kulisch, U.; Miranker, W. L.; Rump, S. M.; Ullrich, Ch.; Wolff v. Gudenberg, J.: *PASCAL–SC: A PASCAL for Contemporary Scientific Computation.* IBM Research Report RC 9009 (#39456) 8/25/81, 79 pages, 1981.
18. Bohlender, G.; Kaucher, E.; Klatte, R.; Kulisch, U.; Miranker, W. L.; Ullrich, Ch.; Wolff v. Gudenberg, J.: *FORTRAN for Contemporary Numerical Computation.* IBM Research Report RC 8348. Computing **26**, pp. 277-314, 1981.
19. Bohlender, G.: *What Do We Need Beyond IEEE Arithmetic?* In [94, pp. 1-32], 1990.
20. Bohlender, G.: *Literature List on Enclosure Methods and Related Topics* Institut für Angewandte Mathematik, Universität Karlsruhe, Report, 1998.
21. Böhm, H.: *Berechnung von Polynomnullstellen und Auswertung arithmetischer Ausdrücke mit garantierter maximaler Genauigkeit.* Dissertation, Universität Karlsruhe, 1983.
22. Braune, K: *Hochgenaue Standardfunktionen für reelle und komplexe Punkte und Intervalle in beliebigen Gleitpunktrastern.* Dissertation, Universität Karlsruhe, 1987.
23. Cappello, P. R.; Miranker, W. L.: *Systolic Super Summation.* IEEE Transactions on Computers **37** (6), pp. 657-677, June 1988.
24. Cappello, P. R.; Miranker, W. L.: *Systolic Super Summation with Reduced Hardware.* IBM Research Report RC 14259 (#63831), IBM Research Division, Yorktown Heights, New York, Nov. 30, 1988.
25. Dietrich, St.: *Adaptive verifizierte Lösung gewöhnlicher Differentialgleichungen.* Dissertation, Universität Karlsruhe, 2002.
26. Erb, H.: *Ein Gleitpunkt-Arithmetikprozessor mit mehrfacher Präzision zur verifizierten Lösung linearer Gleichungssysteme.* Dissertation, Fakultät für Informatik, Universität Karlsruhe, 1992.
27. Facius, A.: *Iterative Solution of Linear Systems with Improved Arithmetic and Result Verification.* Dissertation Universität Karlsruhe, 2000.
28. Facius, A.: *Highly Accurate Verified Error Bounds for Krylov Type Linear System Solvers.* pp. 76-98, in [71].
29. Fischer, H.-C.: *Schnelle Automatische Differentiation, Einschließungsmethoden und Anwendungen.* Dissertation, Universität Karlsruhe, 1990.
30. Fischer, H.-C.: *Automatic Differentiation and Applications.* pp. 105-142, in [1].
31. Hamada, H.: *A New Real Number Representation and its Operation.* In [105, Vol. 8, pp. 153-157], 1987.
32. Hammer, R.: *How Reliable is the Arithmetic of Vector Computers.* pp. 467-482, in [95]., 1990.
33. Hammer, R.: *Maximal genaue Berechnung von Skalarproduktausdrücken und hochgenaue Auswertung von Programmteilen.* Dissertation, Universität Karlsruhe, 1992.
34. Hammer, R.; Hocks, M.; Kulisch, U.; Ratz, D.: **Numerical Toolbox for Verified Computing I: Basic Numerical Problems.** (Vol. II see [57], version in C++ see [35]) Springer–Verlag, Berlin / Heidelberg / New York, 1993.
35. Hammer, R.; Hocks, M.; Kulisch, U.; Ratz, D.: **C++ Toolbox for Verified Computing: Basic Numerical Problems.** Springer–Verlag, Berlin / Heidelberg / New York, 1995.
36. Hergenhan, A.: *Spezifikation und Entwurf einer hochleistungsfähigen Gleitkomma–Architektur.* Diplomarbeit, Technische Universität Dresden, 1994.

37. Hoefflinger, B.: *Next-Generation Floating-Point Arithmetic for Top-Performance PCs.* The 1995 Silicon Valley Personal Computer Design Conference and Exposition, Conference Proceedings, pp. 319-325, 1995.
38. Hoff, T.: *How Children Accumulate Numbers or Why We Need a Fifth Floating-Point Operation.* In: Jahrbuch Überblicke Mathematik, S. 219-222, Vieweg Verlag, 1993.
39. Hofschuster, W.: *Zur Berechnung von Funktionseinschließungen bei speziellen Funktionen der mathematischen Physik.* Dissertation, Universität Karlsruhe, 2000.
40. Hofschuster, W.; Krämer, W.: *A Computer Oriented Approach to Get Sharp Reliable Error Bounds.* Reliable Computing, Issue 3, Volume 3, 1997.
41. Januschke, P.: *Oberon-XSC, Eine Programmiersprache und Arithmetikbibliothek für das Wissenschaftliche Rechnen.* Dissertation, Universität Karlsruhe, 1998.
42. Kelch, R.: *Ein adaptives Verfahren zur numerischen Quadratur mit automatischer Ergebnisverifikation.* Dissertation, Univeristät Karlsruhe, 1989.
43. Kernhof, J.; Baumhof, Ch.; Höfflinger, B.; Kulisch, U.; Kwee, S.; Schramm, P.; Selzer, M.; Teufel, Th.: *A CMOS Floating-Point Processing Chip for Verified Exact Vector Arithmetic.* European Solid State Circuits Conference 94 ESSCIRC, Ulm, Sept. 1994.
44. Kirchner, R.; Kulisch, U.: *Arithmetic for Vector Processors.* In [105, Vol. 8, pp. 256-269], 1987.
45. Kirchner, R.; Kulisch, U.: *Accurate Arithmetic for Vector Processing.* Journal of Parallel and Distributed Computing **5**, special issue on "High Speed Computer Arithmetic", pp. 250-270, 1988.
46. Klatte, R.; Kulisch, U.; Neaga, M.; Ratz, D.; Ullrich, Ch.: **PASCAL–XSC — Sprachbeschreibung mit Beispielen.** Springer-Verlag, Berlin/Heidelberg/New York, 1991 (ISBN 3-540-53714-7, 0-387-53714-7).
47. Klatte, R.; Kulisch, U.; Neaga, M.; Ratz, D.; Ullrich, Ch.: **PASCAL–XSC — Language Reference with Examples.** Springer-Verlag, Berlin/Heidelberg/New York, 1992.
48. Klatte, R.; Kulisch, U.; Lawo, C.; Rauch, M.; Wiethoff, A.: **C–XSC, A C++ Class Library for Extended Scientific Computing.** Springer-Verlag, Berlin/Heidelberg/New York, 1993.
49. Klatte, R.; Kulisch, U.; Neaga, M.; Ratz, D.; Ullrich, Ch.: **PASCAL–XSC — Language Reference with Examples (In Russian).** Moscow, 1994.
50. Knöfel, A.: *Hardwareentwurf eines Rechenwerks für semimorphe Skalar- und Vektoroperationen unter Berücksichtigung der Anforderungen verifizierender Algorithmen.* Dissertation, Universität Karlsruhe, 1991.
51. Klein, W.: *Zur Einschließung der Lösung von linearen und nichtlinearen Fredholmschen Integralgleichungssystemen zweiter Art.* Dissertation, Universität Karlsruhe, 1990.
52. Knöfel, A.: *Fast Hardware Units for the Computation of Accurate Dot Products.* In [105, Vol. 10, pp. 70-74], 1991.
53. Krämer, W.: *Inverse Standardfunktionen für reelle und komplexe Intervallargumente mit a priori Fehlerabschätzungen für beliebige Datenformate.* Dissertation, Universität Karlsruhe, 1987.
54. Krämer, W.: *Constructive Error Analysis.* Journal of Universal Computer Science (JUCS), Vol. 4, No.2, pp. 147-163, 1998.
55. Krämer, W.; Bantle, A.: *Automatic Forward Error Analysis for Floating-Point Algorithms.* Reliable computing, Vol. 7, No. 4, pp. 321-340, 2001.

56. Krämer, W.; Walter, W.: *FORTRAN–SC: A FORTRAN Extension for Engineering/Scientific Computation with Access to ACRITH, General Information Notes and Sample Programs.* pp 1–51, IBM Deutschland GmbH, Stuttgart, 1989.
57. Krämer, W.; Kulisch, U.; Lohner, R.: **Numerical Toolbox for Verified Computing II: Theory, Algorithms and Pascal-XSC Programs.** (Vol. I see [34, 35]) Springer–Verlag, Berlin / Heidelberg / New York, to appear.
58. Kulisch, U.: *An axiomatic approach to rounded computations.* TS Report No. 1020, Mathematics Research Center, University of Wisconsin, Madison, Wisconsin, 1969, and Numerische Mathematik **19**, pp. 1-17, 1971.
59. Kulisch, U.: *Formalization and Implementation of Floating-Point Arithmetic.* Computing **14**, pp. 323-348, 1975.
60. Kulisch, U.: **Grundlagen des Numerischen Rechnens — Mathematische Begründung der Rechnerarithmetik.** Reihe Informatik, Band 19, Bibliographisches Institut, Mannheim/Wien/Zürich, 1976 (ISBN 3-411-01517-9).
61. Kulisch, U.: *Schaltungsanordnung und Verfahren zur Bildung von Skalarprodukten und Summen von Gleitkommazahlen mit maximaler Genauigkeit.* Patentschrift DE 3144015 A1, 1981.
62. Kulisch, U.; Miranker, W. L.: **Computer Arithmetic in Theory and Practice.** Academic Press, New York, 1981 (ISBN 0-12-428650-x).
63. Kulisch, U.; Ullrich, Ch. (Eds.): **Wissenschaftliches Rechnen und Programmiersprachen.** Proceedings of Seminar held in Karlsruhe, April 2–3, 1982. Berichte des German Chapter of the ACM, Band 10, B. G. Teubner Verlag, Stuttgart, 1982 (ISBN 3-519-02429-2).
64. Kulisch, U.; Miranker, W. L. (Eds.): **A New Approach to Scientific Computation.** Proceedings of Symposium held at IBM Research Center, Yorktown Heights, N. Y., 1982. Academic Press, New York, 1983 (ISBN 0-12-428660-7).
65. Kulisch, U.; Miranker, W. L.: *The Arithmetic of the Digital Computer: A New Approach.* IBM Research Center RC 10580, pp. 1-62, 1984. SIAM Review, Vol. 28, No. 1, pp. 1-40, March 1986.
66. Kulisch, U.; Kirchner, R.: *Schaltungsanordnung zur Bildung von Produktsummen in Gleitkommadarstellung, insbes. von Skalarprodukten.* Patentschrift DE 3703440 C2, 1986.
67. Kulisch, U. (Ed.): **PASCAL–SC: A PASCAL extension for scientific computation**, Information Manual and Floppy Disks, Version IBM PC/AT; Operating System DOS. B. G. Teubner Verlag (Wiley-Teubner series in computer science), Stuttgart, 1987 (ISBN 3-519-02106-4 / 0-471-91514-9).
68. Kulisch, U. (Ed.): **PASCAL–SC: A PASCAL extension for scientific computation**, Information Manual and Floppy Disks, Version ATARI ST. B. G. Teubner Verlag, Stuttgart, 1987 (ISBN 3-519-02108-0).
69. Kulisch, U. (Ed.): **Wissenschaftliches Rechnen mit Ergebnisverifikation — Eine Einführung.** Ausgearbeitet von S. Geörg, R. Hammer und D. Ratz. Vol. 58. Akademie Verlag, Berlin, und Vieweg Verlagsgesellschaft, Wiesbaden, 1989.
70. Kulisch, U.; Teufel, T.; Hoefflinger, B.: *Genauer und trotzdem schneller, Ein neuer Coprozessor für hochgenaue Matrix- und Vektoroperationen.* Titelgeschichte, Elektronik **26**, 1994.
71. Kulisch, U.; Lohner, R. and Facius, A. (edts.): **Perspectives on Enclosure Methods.** Springer-Verlag, Wien, New York, 2001.
72. Lichter, P.: *Realisierung eines VLSI-Chips für das Gleitkomma-Skalarprodukt der Kulisch-Arithmetik.* Diplomarbeit, Fachbereich 10, Angewandte Mathematik und Informatik, Universität des Saarlandes, 1988.

73. Meis, T.: *Brauchen wir eine Hochgenauigkeitsarithmetik?* Elektronische Rechenanlagen, Carl Hanser Verlag, pp. 19-23, 1987.
74. Müller, M.; Rüb, Ch.; Rülling, W.: *Exact Accumulation of Floating-Point Numbers.* In [105, Vol. 10, pp. 64-69], 1991.
75. Müller, M.: *Entwicklung eines Chips für auslöschungsfreie Summation von Gleitkommazahlen.* Dissertation, Universität des Saarlandes, Saarbrücken, 1993.
76. Pichat, M.: *Correction d'une somme en arithmétique à virgule flottante.* Numerische Mathematik **19**, pp. 400-406, 1972.
77. Priest, D. M.: *Algorithms for Arbitrary Precision Floating Point Arithmetic.* In [105, Vol. 10, pp. 132-143], 1991.
78. Ratz, D.: *The Effects of the Arithmetic of Vector Computers on Basic Numerical Methods.* pp. 499-514, in [95]., 1990.
79. Ratz, D.: *Automatische Ergebnisverifikation bei globalen Optimierungsproblemen.* Dissertation, Universität Karlsruhe, 1992.
80. Ratz, D.: **Automatic Slope Computation and its Application in Nonsmooth Global Optimization.** Shaker-Verlag, Aachen, 1998.
81. Ratz, D.: *Nonsmooth Global Optimization.* pp. 277-338, in [71].
82. Rojas, R.: *Die Architektur der Rechenmaschinen Z1 und Z3 von Konrad Zuse.* Informatik Spektrum 19/6, Springer-Verlag, pp. 303-315, 1996.
83. Rump, S. M.: *Kleine Fehlerschranken bei Matrixproblemen.* Dissertation, Universität Karlsruhe, 1980.
84. Rump, S. M.: *How Reliable are Results of Computers? / Wie zuverlässig sind die Ergebnisse unserer Rechenanlagen?* In: *Jahrbuch Überblicke Mathematik*, Bibliographisches Institut, Mannheim, 1983.
85. Rump, S. M.: *Solving algebraic problems with high accuracy.* pp.51-120, in [64]., 1983.
86. Rump, S. M.; Böhm, H.: *Least Significant Bit Evaluation of Arithmetic Expressions in Single-Precision.* Computing **30**, pp. 189-199, 1983.
87. Schmidt, L.: *Semimorphe Arithmetik zur automatischen Ergebnisverifikation auf Vektorrechnern.* Dissertation, Universität Karlsruhe, 1992.
88. Shiriaev, D.: *Fast Automatic Differentiation for Vector Processors and Reduction of the Spatial Complexity in a Source Translation Environment.* Dissertation, Universität Kalrsuhe, 1993.
89. Stetter, H. J.: *Sequential Defect Correction for High-Accuracy Floating-Point Algorithms.* Lecture Notes in Mathematics, Vol. 1006, pp. 186-202, Springer-Verlag, 1984.
90. Stetter, H. J.: *Staggered Correction Representation, a Feasible Approach to Dynamic Precision.* In: *Proceedings of the Symposium on Scientific Software*, edited by Cai, Fosdick, Huang, China University of Science and Technology Press, Beijing, China, 1989.
91. Suzuki, H.; Morinaka, H.; Makino, H.; Nakase, Y.; Mashiko, K.; Sumi, T,: *Leading-Zero Anticipatory Logic for High-Speed Floating-Point Addition.* IEEE Journal of Solid-State Circuits, Vol. 31, No. 8, August 1996.
92. Tangelder, R.J.W.T: *The Design of Chip Architectures for Accurate Inner Product Computation.* Dissertation, Technical University Eindhoven, 1992. ISBN 90-9005204-6.
93. Teufel, T.: *Ein optimaler Gleitkommaprozessor.* Dissertation, Universität Karlsruhe, 1984.
94. Ullrich, Ch. (Ed.): **Computer Arithmetic and Self-Validating Numerical Methods.** (Proceedings of SCAN 89, held in Basel, Oct. 2-6, 1989, invited papers). Academic Press, San Diego, 1990.

95. Ullrich, Ch. (Ed.): **Contributions to Computer Arithmetic and Self-Validating Numerical Methods.** J.C.Baltzer AG, Scientific Publishing Co., 1990.
96. Wallis, P. J. L. (Ed.): **Improving Floating-Point Programming.** J. Wiley, Chichester, 1990 (ISBN 0 471 92437 7).
97. Walter, W.: *FORTRAN–SC: A FORTRAN Extension for Engineering / Scientific Computation with Access to ACRITH, Language Reference and User's Guide.* 2nd ed., pp. 1-396, IBM Deutschland GmbH, Stuttgart, Jan. 1989.
98. Walter, W.: *Flexible Precision Control and Dynamic Data Structures for Programming Mathematical and Numerical Algorithms.* 1990.
99. Walter, W. V.: *Mathematical Foundations of Fully Reliable and Portable Software for Scientific Computing.* Universität Karlsruhe, 1995.
100. Wilkinson, J.: **Rounding Errors in Algebraic Processes.** Prentice-Hall, Englewood Cliffs, New Jersey, 1963.
101. Winter, Th.: *Ein VLSI-Chip für Gleitkomma-Skalarprodukt mit maximaler Genauigkeit.* Diplomarbeit, Fachbereich 10, Angewandte Mathematik und Informatik, Universität des Saarlandes, 1985.
102. Winter, D. T.: *Automatic Identification of Scalar Products.* In [96], 1990.
103. Yilmaz, T.; Theeuwen, J.F.M.; Tangelder, R.J.W.T.; Jess, J.A.G.: *The Design of a Chip for Scientific Computation.* Eindhoven University of Technology, 1989 and pp. 335-346 of Proceedings of the Euro-Asic Symposium, Grenoble, Jan.25-27, 1989.
104. Yohe, J.M.: *Roundings in Floating-Point Arithmetic.* IEEE Trans. on Computers, Vol. C-22, No. 6, June 1973, pp. 577-586.
105. Institute of Electrical and Electronics Engineers: **Proceedings of x-th Symposium on Computer Arithmetic ARITH.** IEEE Computer Society Press. IEEE Service Center, 445 Hoes Lane, P.O.Box 1331, Piscataway, NJ 08855-1331, USA.
    Editors of proceedings; place of conference; date of conference.
    1. Shively, R.R.; Minneapolis; June 16, 1969.
    2. Garner, H.L.; Atkins, D.E.; Univ Maryland, College Park; May 15 – 16, 1972.
    3. Rao, T.R.N.; Matula, D.W.; SMU, Dallas; Nov. 19 – 20, 1975.
    4. Avizienis, A.; Ercegovac, M.D.; UCLA, Los Angeles; Oct. 25 – 27, 1978.
    5. Trivedi, K.S.; Atkins, D.E.; Univ Michigan, Ann Arbor; May 18 – 19, 1981.
    6. Rao, T.R.N.; Kornerup, P.; Univ Aarhus, Denmark; June 20 – 22, 1983.
    7. Hwang, K.; Univ Illinois, Urbana; June 4 – 6, 1985.
    8. Irwin, M.J.; Stefanelli, R.; Como, Italy; May 19 – 21, 1987.
    9. Ercegovac, M.; Swartzlander, E.; Santa Monica; Sept. 6 – 8, 1989.
    10. Kornerup, P.; Matula, D.; Grenoble, France; June 26 – 28, 1991.
    11. Swartzlander Jr., E.; Irwin, M. J.; Jullien, G.; Windsor, Ontario; June 29 – July 2, 1993.
    12. Knowles, S.; Mc Allister, W. H.; Bath, England; July 19 – 21, 1995;
    13. Lang, Th.; Muller, J.-M.; Takagi, N.; Asilomar, California; July 6 – 9, 1997;
106. IAM: *PASCAL-XR: PASCAL for eXtended Real arithmetic.* Joint research project with Nixdorf Computer AG. Institute of Applied Mathematics, University of Karlsruhe, Postfach 6980, D-76128 Karlsruhe, Germany, 1980.
107. IAM: *FORTRAN–SC: A FORTRAN Extension for Engineering / Scientific Computation with Access to ACRITH.* Institute of Applied Mathematics, University of Karlsruhe, Postfach 6980, D-76128 Karlsruhe, Germany, Jan. 1989.
    1. Language Reference and User's Guide, 2nd edition.
    2. General Information Notes and Sample Programs.

108. IAM: *ACRITH–XSC, A Programming Language for Scientific Computation.* Syntax Diagrams. Institute of Applied Mathematics, University of Karlsruhe, Postfach 6980, D-76128 Karlsruhe, Germany, 1990.
109. IBM: *IBM System/370 RPQ. High Accuracy Arithmetic.* SA 22-7093-0, IBM Deutschland GmbH (Department 3282, Schönaicher Strasse 220, D-71032 Böblingen), 1984.
110. IBM: **IBM High-Accuracy Arithmetic Subroutine Library (ACRITH).** IBM Deutschland GmbH (Department 3282, Schönaicher Strasse 220, D-71032 Böblingen), 3rd edition, 1986.
    1. General Information Manual. GC 33-6163-02.
    2. Program Description and User's Guide. SC 33-6164-02.
    3. Reference Summary. GX 33-9009-02.
111. IBM *Verfahren und Schaltungsanordnung zur Addition von Gleitkommazahlen.* Europäische Patentanmeldung, EP 0 265 555 A1, 1986.
112. IBM: **ACRITH–XSC: IBM High Accuracy Arithmetic — Extended Scientific Computation. Version 1, Release 1.** IBM Deutschland GmbH (Schönaicher Strasse 220, D-71032 Böblingen), 1990.
    1. General Information, GC33-6461-01.
    2. Reference, SC33-6462-00.
    3. Sample Programs, SC33-6463-00.
    4. How To Use, SC33-6464-00.
    5. Syntax Diagrams, SC33-6466-00.
113. IEEE: *A Proposed Standard for Binary Floating-Point Arithmetic.* IEEE Computer, March 1981.
114. American National Standards Institute / Institute of Electrical and Electronics Engineers: *A Standard for Binary Floating-Point Arithmetic.* ANSI/IEEE Std. 754-1985, New York, 1985 (reprinted in SIGPLAN **22**, 2, pp. 9-25, 1987). Also taken over as IEC Standard 559:1989.
115. American National Standards Institute / Institute of Electrical and Electronics Engineers: *A Standard for Radix-Independent Floating-Point Arithmetic.* ANSI/IEEE Std. 854-1987, New York, 1987.
116. IMACS; GAMM: *IMACS-GAMM Resolution on Computer Arithmetic.* In Mathematics and Computers in Simulation **31**, pp. 297-298, 1989. In Zeitschrift für Angewandte Mathematik und Mechanik **70**, no. 4, p. T5, 1990.
117. IMACS; GAMM: *GAMM-IMACS Proposal for Accurate Floating-Point Vector Arithmetic.* GAMM, Rundbrief 2, pp. 9-16, 1993. Mathematics and Computers in Simulation, Vol. **35**, IMACS, North Holland, 1993. News of IMACS, Vol. 35, No. 4, pp. 375-382, Oct. 1993.
118. Numerik Software GmbH: **PASCAL–XSC: A PASCAL Extension for Scientific Computation. User's Guide.** Numerik Software GmbH, Haid- und-Neu-Straße 7, D-76131 Karlsruhe, Germany / Postfach 2232, D-76492 Baden-Baden, Germany, 1991.
119. SIEMENS: **ARITHMOS (BS 2000) Unterprogrammbibliothek für Hochpräzisionsarithmetik. Kurzbeschreibung, Tabellenheft, Benutzerhandbuch.** SIEMENS AG, Bereich Datentechnik, Postfach 83 09 51, D-8000 München 83. Bestellnummer U2900-J-Z87-1, Sept. 1986.

2. Rounding Near Zero

Summary.
This paper deals with arithmetic on a discrete subset S of the real numbers $I\!R$ and with floating-point arithmetic in particular. We assume that arithmetic on S is defined by semimorphism. Then for any element $a \in S$ the element $-a \in S$ is an additive inverse of a, i.e. $a \oplus (-a) = 0$. The first part of the paper describes a necessary and sufficient condition under which $-a$ is the unique additive inverse of a in S. In the second part this result is generalized. We consider algebraic structures M which carry a certain metric, and their semimorphic images on a discrete subset N of M. Again, a necessary and sufficient condition is given under which elements of N have a unique additive inverse. This result can be applied to complex floating-point numbers, real and complex floating-point intervals, real and complex floating-point matrices, and real and complex floating-point interval matrices.

2.1 The one dimensional case

Let $I\!R$ denote the set of real numbers and S a discrete subset of $I\!R$ which is symmetric with respect to zero, i.e. $0 \in S$ and for all $a \in S$ also $-a \in S$. A semimorphism defines arithmetic operations $\oplus$ and $\otimes$ on S by the following rules:

(RG) $\quad a \odot b := \bigcirc (a \circ b) \qquad$ for all $a, b \in S$ and $\circ \in \{+, *\}$.

In (RG) $\bigcirc$ is a rounding $\bigcirc : I\!R \to S$ with the properties

(R1) $\quad \bigcirc(a) = a \qquad$ for all $a \in S \qquad$ (projection).
(R2) $\quad a \leq b \Rightarrow \bigcirc(a) \leq \bigcirc(b) \qquad$ for $a, b \in I\!R \qquad$ (monotonicity).
(R3) $\quad \bigcirc(-a) = -\bigcirc(a) \qquad$ for all $a \in I\!R \qquad$ (antisymmetry).

That is, the rounding is a monotone and antisymmetric projection of $I\!R$ onto S.

A semimorphism leads to the best possible arithmetic in S in the sense that between the computed result of an operation and the correct result there is never another element of S. The computed result is rounding dependent.

In case of rounding to nearest it is accurate to 1/2 unit in the last place (ulp). In all other cases it is accurate to 1 ulp.

Typical monotone and antisymmetric roundings are the rounding to the nearest element in S, the rounding towards zero, and the rounding away from zero.

Since $0 \in S$, (R1) and (RG) yield immediately that for every $a \in S$ the element $-a$ is an additive inverse of a in S:

$$a \oplus (-a) = \bigcirc(a + (-a)) = \bigcirc(0) = 0 \text{ for all } a \in S.$$

However, in a normalized floating-point system $-a$ is not in general the only additive inverse of a in S. We briefly recall the definition of normalized floating-point numbers:

A normalized floating-point number is a real number of the form:

$$x = \circ m \cdot b^e.$$

Here $\circ \in \{+, -\}$ is the sign of the number, m is the mantissa, b is the base of the number system in use and e is the exponent. b is an integer greater than unity. The exponent is an integer between two fixed integer bounds e_1, e_2, and in general $e_1 \leq 0 \leq e_2$. The mantissa is of the form

$$m = \sum_{i=1}^{r} d_i \cdot b^{-i}.$$

The d_i are the digits of the mantissa. They have the property $d_i \in \{0, 1, \ldots, b-1\}$ for all $i = 1(1)r$ and $d_1 \neq 0$. Without the condition $d_1 \neq 0$, floating-point numbers are said to be unnormalized. The set of normalized floating-point numbers does not contain zero. So zero is adjoined to S. For a unique representation of zero it is often assumed that $m = 0.00 \cdots 0$ and $e = 0$. A floating-point system depends on the constants b, r, e_1, and e_2. We denote it by $S = S(b, r, e_1, e_2)$.

The floating-point numbers are not equally spaced between successive powers of b and their negatives. This spacing changes at every power of b. In particular, there are relatively large gaps around zero which contain no further floating-point number. Figure 2.1 shows a simple floating-point system $S = S(2, 3, -1, 2)$ consisting of 33 elements.

If, for instance, the rounding towards zero is chosen, the entire interval $(-1/4, 1/4)$ is mapped onto zero. So whenever the real sum of two numbers of S falls into this interval (e.g. $1/4 - 3/8$) their sum in S is zero, $a \oplus b = 0$, and the two elements form a pair of additive inverses.

The following theorem characterizes a discrete subset S of $\mathbb{R}$ by a necessary and sufficient condition under which the element $-a$ is the unique additive inverse of a in S:

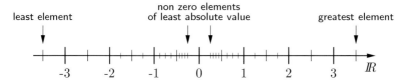

Fig. 2.1. The characteristic spacing of a floating-point system.

Theorem 1:
If S is a symmetric, discrete subset of $\mathbb{R}$ with $0 \in S$, $\bigcirc: \mathbb{R} \to S$ a semimorphism, and $\varepsilon > 0$ the least distance between distinct elements of S, then for all $a \in S$ the element $b = -a$ is the unique additive inverse of a if and only if
$$\bigcirc^{-1}(0) \subseteq (-\varepsilon, \varepsilon). \tag{2.1}$$

Here $\bigcirc^{-1}(0)$ denotes the inverse image of 0 and $(-\varepsilon, \varepsilon)$ is the open interval between $-\varepsilon$ and ε.

Proof. At first we show that (2.1) is sufficient: We assume that $b \neq -a$ is an additive inverse of a in S. Then $a \oplus b = 0$ and by (RG) $a + b \in \bigcirc^{-1}(0)$. This means by (2.1)
$$-\varepsilon < a + b < \varepsilon. \tag{2.2}$$

Since $-b \neq a$ we have by definition of ε: $|a - (-b)| = |a + b| \geq \varepsilon$. This contradicts (2.2), so the assumption is false. Under the condition (2.1) there is no additive inverse of a other than $-a$. In other words (2.1) is sufficient.

Now we show that (2.1) is necessary also: Since $0 \in S$ by (R1) $\bigcirc(0) = 0$ and $0 \in \bigcirc^{-1}(0)$. Since $\bigcirc$ is monotone, $\bigcirc^{-1}(0)$ is convex. Since $\bigcirc$ is antisymmetric, $\bigcirc^{-1}(0)$ is symmetric with respect to zero. We assume now that (2.1) is not true, then $\bigcirc^{-1}(0) \supset (-\varepsilon, \varepsilon)$. We take two elements $a, b \in S$ with distance ε. Then $a \neq b$ and $|a - b| = \varepsilon$, i.e. $a + (-b) \in \bigcirc^{-1}(0)$ or $a \oplus (-b) = 0$. This means that $-b \neq -a$ is inverse to a. Thus (2.1) is necessary because otherwise there would be more than one additive inverse to a. □

(2.1) holds automatically if the number ε itself is an element of S. Then, because of (R1), $\bigcirc(\varepsilon) = \varepsilon$ and, because of the monotonicity of the rounding (R2) $\bigcirc^{-1}(0) \subseteq (-\varepsilon, \varepsilon)$. In other words: if the least distance ε of two elements of a discrete subset S of $\mathbb{R}$ is an element of S, then for all $a \in S$ the element $-a \in S$ is the unique additive inverse of a in S. (This holds under the assumption that the mapping $\bigcirc: \mathbb{R} \to S$ is a semimorphism).

Such subsets of the real numbers do indeed occur. For instance, the integers or the so called fixed-point numbers are subsets of $\mathbb{R}$ with this property. Sometimes the normalized floating-point numbers are extended into a set with this property. Such a set is obtained if in the case $e = e_1$ unnormalized

74   2. Rounding Near Zero

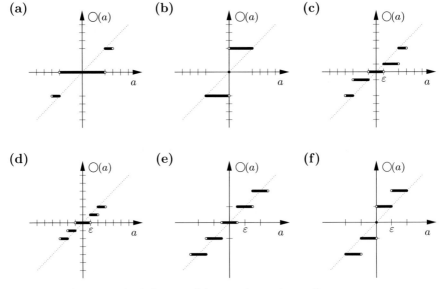

Fig. 2.2. The behavior of frequently used roundings near zero.

mantissas are permitted. Then ε itself becomes an element of S and for all $a \in S$ the element $-a$ is the unique additive inverse of a. This is the case, for instance, if IEEE arithmetic with denormalized numbers is implemented.

Figure 2.2 illustrates the behavior of typical roundings in the neighborhood of zero. (R1) means that for floating-point numbers the rounding function coincides with the identity mapping.

(a) shows the conventional behavior of the rounding in case of normalized floating-point numbers. In this case (2.1) does not hold and we have no uniqueness of the additive inverse.
(b) shows the rounding away form zero in case of normalized floating-point numbers. In this case (2.1) holds and we have unique additive inverses.
(c) here and in the following cases ε is an element of S and we have unique additive inverses.
(d) shows the rounding toward zero near zero in the case where denormalized numbers are permitted for $e = e_1$. In the IEEE arithmetic standard this situation is called gradual underflow.
(e) shows the rounding to nearest in the neighborhood of zero. The roundings (d) and(e) are provided by the IEEE arithmetic standard.
(f) shows the rounding away from zero in the neighborhood of zero with denormalized numbers permitted for $e = e_1$. This rounding has all required properties. It is $\bigcirc(a) = 0$ if and only if $a = 0$, a property which can be very important for making a clear distinction between a number that is

zero and a number that is not, however small it might actually be. This rounding is not provided by the IEEE arithmetic standard.

2.2 Rounding in product spaces

In addition to the real numbers, other spaces frequently occur in numerical mathematics. Such spaces are the complex numbers, the real and complex intervals, the real and complex matrices, and the real and complex interval matrices. In their computer representable subsets subtraction is no longer the inversion of addition nor is division the inversion of multiplication. Nevertheless subtraction is not an independent operation. If arithmetic in the computer representable subspaces is defined by semimorphism, subtraction can be defined by addition and multiplication with the negated multiplicative unit. For clarity we briefly show the definition here.

Let M denote any one of the sets listed above and N its computer representable subset. A semimorphism defines arithmetic operations $\oplus$ and $\odot$ in N by

(RG) $\quad a \odot b := \bigcirc(a \circ b) \qquad$ for all $a, b \in N$ and $\circ \in \{+, *\}$.

Here $\bigcirc : M \to N$ is a mapping with the following properties

(R1) $\quad \bigcirc(a) = a \qquad$ for all $a \in N \qquad$ (rounding).
(R2) $\quad a \leq b \Rightarrow \bigcirc(a) \leq \bigcirc(b) \qquad$ for $a, b \in M \qquad$ (monotonicity).
(R3) $\quad \bigcirc(-a) = -\bigcirc(a) \qquad$ for all $a \in M \qquad$ (antisymmetry).

In case of the interval spaces, the order relation $\leq$ means set inclusion $\subseteq$. In this case the rounding is required to have the additional property

(R4) $\quad a \leq \bigcirc(a) \qquad$ for all $a \in M \qquad$ (upwardly directed).

With this definition it is shown in [3, 4] for all spaces mentioned above that the multiplicative unit e has a unique additive inverse $\ominus e$ in N. With this quantity the minus operator (negation) and subtraction are defined by

$$\ominus b := (\ominus e) \odot b \quad \text{and} \quad a \ominus b := a \oplus (\ominus b). \qquad (2.3)$$

This preserves all rules of the minus operator in the computer representable subspaces, [3, 4].

The proof that $\ominus e$ is unique is intricate and not easy in all the individual cases [3, 4]. A generalization of the conditions given in Section 2.1 for the

2. Rounding Near Zero

existence and uniqueness of additive inverses for the product spaces listed above could simplify the situation considerably. This is now done.

We assume that the basic set M is mapped into a discrete subset N by semimorphism, where N is symmetric, i.e. $0 \in N$ and for all $a \in N$ also $-a \in N$. It follows from (RG) and (R1) that an element $a \in N$ which has an additive inverse $-a$ in M has the same additive inverse in N:

$$a \oplus (-a) := \bigcirc(a + (-a)) = \bigcirc(0) = 0.$$

Now we assume additionally that M is a metric space with a distance function $d: M \times M \to \mathbb{R}$ which has the property:

$$d(a+c, b+c) = d(a,b) \text{ for all } a, b, c \in M \quad \text{(translation invariant)}. \quad (2.4)$$

With d, the property of N being discrete can now be expressed by

$$d(a,b) \geq \varepsilon > 0 \qquad \text{for all } a, b \in N \text{ with } a \neq b, \quad (2.5)$$

where $\varepsilon > 0$ is the least distance of distinct elements of N. With these concepts the following theorem holds:

Theorem 2:
For all elements a of N which have an unique additive inverse $-a$ in M, $-a$ is also the unique additive inverse of a in N if and only if

$$d(x, 0) < \varepsilon \qquad \text{for all } x \in M \text{ with } \bigcirc(x) = 0. \quad (2.6)$$

Proof. At first we show that (2.6) is sufficient: We assume that $b \neq -a$ is an additive inverse of a in N. Then we obtain by (RG) and (2.6):

$$a \oplus b = 0 \quad \Rightarrow \quad \bigcirc(a+b) = 0 \quad \Rightarrow \quad d(a+b, 0) < \varepsilon. \quad (2.7)$$

On the other hand we get by the definition of ε and by (2.4):

$$d(b, -a) \geq \varepsilon \quad \Rightarrow \quad d(a+b, a+(-a)) = d(a+b, 0) \geq \varepsilon.$$

This contradicts (2.7), so the assumption that there is an additive inverse b of a other than $-a$ is false. In other words (2.6) is sufficient.

Now we show that (2.6) is necessary also: By (R1) we obtain $0 \in \bigcirc^{-1}(0)$. As a consequence of (R2) $\bigcirc^{-1}(0)$ is convex and by (R3) $\bigcirc^{-1}(0)$ is symmetric, since for all $a \in M$

$$a \in \bigcirc^{-1}(0) \quad \Rightarrow \quad \bigcirc(a) = 0 \quad \Rightarrow \quad \bigcirc(-a) = -\bigcirc(a) = 0 \quad \Rightarrow \quad -a \in \bigcirc^{-1}(0).$$

Should (2.6) not hold there would be elements $x \in M$ with

$$d(x, 0) \geq \varepsilon > 0 \text{ and } \bigcirc(x) = 0.$$

ε is the least distance of distinct elements of N. Now we choose two different elements $a, b \in N$ with distance ε, for instance $a = x$ and $b = 0$. Then $d(a, b) = \varepsilon$ and $\bigcirc(a - b) = 0$, i.e. $a - b \in \bigcirc^{-1}(0)$ or $a \oplus (-b) = 0$. This means that $-b$ is inverse to a and $-b \neq -a$. In other words: if (2.6) does not hold there are elements in N which have more than one additive inverse. This shows that (2.6) is necessary which completes the proof. □

To fully establish the theorem of Section 2.2 we still have to demonstrate that in all the basic spaces under consideration a metric does indeed exist which is translation invariant, see (2.4). We just mention the appropriate metric and leave the demonstration of (2.4) to the reader:

- If M is the set of real numbers $\mathbb{R}$, then $d(a, b) = |a - b|$.
- If M is the set of complex numbers $\mathbb{C}$, the distance of two complex numbers $a = a_1 + ia_2$ and $b = b_1 + ib_2$ is defined by $d(a, b) := |a_1 - b_1| + |a_2 - b_2|$.
- If M is the set of real intervals $I\mathbb{R}$ the distance of two intervals $a = [a_1, a_2]$ and $b = [b_1, b_2]$ is defined by $d(a, b) := \max(|a_1 - b_1|, |a_2 - b_2|)$.
- If M is the set of complex intervals $I\mathbb{C}$ the distance of two complex intervals $a = a_1 + ia_2$ and $b = b_1 + ib_2$ with real intervals a_1, a_2, b_1 and b_2 is defined by $d(a, b) := d(a_1, b_1) + d(a_2, b_2)$.
- In case of two matrices $a = (a_{ij})$ and $b = (b_{ij})$ with components a_{ij}, b_{ij} of $\mathbb{R}, \mathbb{C}, I\mathbb{R}$, or $I\mathbb{C}$, the distance is defined in each case as the maximum of the distances of corresponding matrix components: $d(a, b) := \max(d(a_{ij}, b_{ij}))$.

In all the basic structures under consideration, which are the sets $\mathbb{R}, \mathbb{C}, I\mathbb{R}, I\mathbb{C}$, and the matrices with components of these sets, the multiplicative unit e has a unique additive inverse $-e$. Under the condition (2.1) and (2.6) respectively, therefore, $-e$ is also the unique additive inverse in any discrete computer representable subset. This allows the definition of the minus operator and subtraction in the computer representable subsets as shown in (2.3) with all its consequences (see [3, 4]).

A closer look at interval spaces is particularly interesting. Again we consider the sets $\mathbb{R}$ and $\mathbb{C}$ as well as the matrices $M\mathbb{R}$ and $M\mathbb{C}$ with components of $\mathbb{R}$ and $\mathbb{C}$. All these sets are ordered with respect to the order relation $\leq$. If M denotes any one of these sets the concept of an interval is defined by

$$A = [a_1, a_2] := \{a \in M \mid a_1, a_2 \in M, a_1 \leq a \leq a_2\}.$$

In the set IM of all such intervals arithmetic operations can be defined, see [1, 3, 4]. For all $M \in \{\mathbb{R}, \mathbb{C}, M\mathbb{R}, M\mathbb{C}\}$ the elements of IM in general are not computer representable and the arithmetic operations in IM are not computer executable. Therefore, subsets $N \subseteq M$ of computer representable elements have to be chosen. An interval in IN is defined by

$$A = [a_1, a_2] := \{a \in M \mid a_1, a_2 \in N, a_1 \leq a \leq a_2\}.$$

Arithmetic operations in IN are defined by semimorphism, i.e. by (RG) with the monotone and antisymmetric rounding $\diamondsuit : IM \to IN$ which is upwardly directed. $\Diamond$ is uniquely defined by these properties (R1), (R2), (R3), and (R4). This process leads to computer executable operations in the interval spaces IN of computer representable subsets of $I\!R$, $\mathbb{C}$, $MI\!R$ and $M\mathbb{C}$.

In this treatise the inverse image of zero with respect to a rounding plays a key role. Since the rounding $\Diamond : IM \to IN$ is upwardly directed with respect to set inclusion as an order relation the inverse image of zero $\Diamond^{-1}(0)$ can only be zero itself. Thus the necessary and sufficient criterion (2.6) for the existence of unique additive inverses evidently holds for IN. Among others this establishes the fact with all its consequences that the unit interval $[e, e]$ has a unique additive inverse $[-e, -e]$ in IN for all discrete subsets N of $M \in \{I\!R, \mathbb{C}, MI\!R, M\mathbb{C}\}$.

Bibliography and Related Literature

1. Alefeld, G.; Herzberger, J.: **An Introduction to Interval Computations.** Academic Press, New York, 1983 (ISBN 0-12-049820-0).
2. Kaucher, E.: *Über metrische und algebraische Eigenschaften einiger beim numerischen Rechnen auftretender Räume.* Dissertation, Universität Karlsruhe, 1973.
3. Kulisch, U.: **Grundlagen des Numerischen Rechnens — Mathematische Begründung der Rechnerarithmetik.** Reihe Informatik, Band 19, Bibliographisches Institut, Mannheim/Wien/Zürich, 1976 (ISBN 3-411-01517-9).
4. Kulisch, U.; Miranker, W. L.: **Computer Arithmetic in Theory and Practice.** Academic Press, New York, 1981 (ISBN 0-12-428650-x).
5. Yohe, J.M.: *Roundings in Floating-Point Arithmetic.* IEEE Trans. on Computers, Vol. C-22, No. 6, June 1973, pp. 577-586.
6. American National Standards Institute/Institute of Electrical and Electronic Engineers: *A Standard for Binary Floating-Point Arithmetic.* ANSI/IEEE Std. 754-1985, New York, 1985 (reprinted in SIGPLAN 22, 2 pp. 9-25, 1987). Also taken over as IRC Standard 559:1989.
7. American National Standards Institute/Institute of Electrical and Electronic Engineers: *A Standard for Radix-Independent Floating-Point Arithmetic.* ANSI/IEEE Std. 854-1987, New York, 1987.

3. Interval Arithmetic Revisited

Summary.

This paper deals with interval arithmetic and interval mathematics. Interval mathematics has been developed to a high standard during the last few decades. It provides methods which deliver results with guarantees. However, the arithmetic available on existing processors makes these methods extremely slow. The paper reviews a number of basic methods and techniques of interval mathematics in order to derive and focus on those properties which by today's knowledge could effectively be supported by the computer's hardware, by basic software, and by the programming languages. The paper is not aiming for completeness. Unnecessary mathematical details, formalisms and derivations are left aside whenever possible. Particular emphasis is put on an efficient implementation of interval arithmetic on computers.

Interval arithmetic is introduced as a shorthand notation and automatic calculus to add, subtract, multiply, divide, and otherwise deal with inequalities. Interval operations are also interpreted as special powerset or set operations. The inclusion isotony and the inclusion property are central and important consequences of this property. The basic techniques for enclosing the range of function values by centered forms or by subdivision are discussed. The Interval Newton Method is developed as an always (globally) convergent technique to enclose zeros of functions.

Then extended interval arithmetic is introduced. It allows division by intervals that contain zero and is the basis for the development of the extended Interval Newton Method. This is the major tool for computing enclosures at all zeros of a function or of systems of functions in a given domain. It is also the basic ingredient for many other important applications like global optimization, subdivision in higher dimensional cases or for computing error bounds for the remainder term of definite integrals in more than one variable. We also sketch the techniques of differentiation arithmetic, sometimes called automatic differentiation, for the computation of enclosures of derivatives, of Taylor coefficients, of gradients, of Jacobian or Hessian matrices.

The major final part of the paper is devoted to the question of how interval arithmetic can effectively be provided on computers. This is an essential prerequisite for its superior and fascinating properties to be more widely used in the scientific computing community. With more appropriate processors, rigorous methods based on interval arithmetic could be comparable in speed with today's "approximate" methods. At processor speeds of gigaFLOPS there remains no alternative but to furnish future computers with the capability to control the accuracy of a computation at least to a certain extent.

3.1 Introduction and Historical Remarks

In 1958 the Japanese mathematician Teruo Sunaga published a paper entitled "Theory of an Interval Algebra and its Application to Numerical Analysis" [62]. Sunaga's paper was intended to indicate a method of rigorous error estimation alternative to the methods and ideas developed in J. v. Neumann and H. H. Goldstine's paper on "Numerical Inverting of Matrices of High Order". [48]

Sunaga's paper is not the first one using interval arithmetic in numerical computing. However, several ideas which are standard techniques in interval mathematics today are for the first time mentioned there in rudimentary form. The structure of interval arithmetic is studied in Sunaga's paper. The possibility of enclosing the range of a rational function by interval arithmetic is discussed. The basic idea of what today is called the Interval Newton Method can be found there, and also the methods of obtaining rigorous bounds in the cases of numerical integration of definite integrals or of initial value problems of ordinary differential equations by evaluating the remainder term of the integration routine in interval arithmetic are indicated in Sunaga's paper. Under "Conclusion" Sunaga's paper ends with the statement "that a future problem will be to revise the structure of the automatic digital computer from the standpoint of interval calculus".

Today Interval Analysis or Interval Mathematics appears as a mature mathematical discipline. However, the last statement of Sunaga's paper still describes a "future problem". The present paper is intended to help close this gap.

This paper is supposed to provide an informal, easily readable introduction to basic features, properties and methods of interval arithmetic. In particular it is intended to deepen the understanding and clearly derive those properties of interval arithmetic which should be supported by computer hardware, by basic software, and by programming languages. The paper is not aiming for completeness. Unnecessary mathematical details, formalisms and derivations are put aside, whenever possible.

Interval mathematics has been developed to a high level during the last decades at only a few academic sites. Problem solving routines which deliver validated results are actually available for all the standard problems of numerical analysis. Many applications have been solved using these tools. Since all these solutions are mathematically proven to be correct, interval mathematics has occasionally been called the Mathematical Numerics in contrast to Numerical Mathematics, where results are sometimes merely speculative. Interval mathematics is not a trivial subject which can just be applied naively. It needs education, training and practice. The author is convinced that with the necessary skills interval arithmetic can be useful, and can be successfully applied to any serious scientific computing problem.

In spite of all its advantages it is a fact that interval arithmetic is not widely used in the scientific computing community as a whole. The author sees several reasons for this which should be discussed briefly. A broad understanding of these reasons is an essential prerequisite for further progress.

Forty years of nearly exclusive use of floating-point arithmetic in scientific computing has formed and now dominates our thinking. Interval arithmetic requires a much higher level of abstraction than languages like Fortran-77, Pascal or C provide. If every single interval operation requires a procedure call, the user's energy and attention are forced down to the level of coding, and are dissipated there.

The development and implementation of adequate and powerful programming environments like PASCAL-XSC [17, 26, 27] or ARITH-XSC [77] requires a large body of experienced and devoted scientists (about 20 man years for each) which is not easy to muster. In such environments interval arithmetic, the elementary functions for the data types real and interval, a long real and a long real interval arithmetic including the corresponding elementary functions, vector and matrix arithmetic, differentiation and Taylor arithmetic both for real and interval data are provided by the run time system of the compiler. All operations can be called by the usual mathematical operator symbols and are of maximum accuracy. This releases the user from coding drudgery. This means, for instance, that an enclosure of a high derivative of a function over an interval — needed for step size control and to guarantee the value of a definite integral or a differential equation within close bounds — can be computed by the same notation used to compute the real function value. The compiler interprets the operators according to the type specification of the data. This level of programming is essential indeed. It opens a new era of conceptual thinking for mathematical numerics.

A second reason for the low acceptance of interval arithmetic in the scientific computing community is simply the prejudices which are often the result of superficial experiments. Sentences like the following appear again and again in the literature: "The error bounds are overly conservative; they quickly grow to the computer representation of $[-\infty, +\infty]$", "Interval arithmetic is expensive because it takes twice the storage and at least twice the work of ordinary arithmetic."

Such sentences are correct for what is called "naive interval arithmetic". Interval arithmetic, however, should not be applied naively. Its properties must be studied and understood first, before it can be applied successfully. Many program packages have been developed using interval arithmetic, which deliver close bounds for their solutions. In no case are these bounds obtained by substituting intervals in a conventional floating-point algorithm. Interval arithmetic is an extension of floating-point arithmetic, not a replacement for it. Sophisticated use of interval arithmetic often leads to safe and better results. There are many applications where the extended tool delivers a guaranteed answer faster than the restricted tool of floating-point arithmetic delivers an "approximation". Examples are numerical integration (because of automatic step size control) and global optimization (intervals bring the continuum on the computer). One interval evaluation of a function over an interval may suffice to prove that the function definitively has no zero in that interval, while 1000 floating-point evaluations of the function in the interval could not provide a safe answer. Interval methods that have been developed for systems of ordinary differential and integral equations may be a bit slower. But they deliver not just unproven numbers. Interval methods deliver close bounds and prove existence and uniqueness of the solution within the computed bounds. The bounds include both discretization and rounding errors. This can save a lot of computing time by avoiding experimental reruns.

The main reason why interval methods are sometimes slow is already expressed in the last statement of Sunaga's early article. It's not that the methods are slow. It is the missing hardware support which makes them slow. While conventional floating-point arithmetic nowadays is provided by fast hardware, interval arithmetic has to be simulated by software routines based on integer arithmetic. The IEEE arithmetic standard, adopted in 1985, seems to support interval arithmetic. It requires the basic four arithmetic operations with rounding to nearest, towards zero, and with rounding downwards and upwards. The latter two are needed for interval arithmetic. But processors that provide IEEE arithmetic separate the rounding from the operation, which proves to be a severe drawback. In a conventional floating-point computation this does not cause any difficulties. The rounding mode is set only once. Then a large number of operations is performed with this rounding mode. However, when interval arithmetic is performed the rounding mode has to be switched very frequently. The lower bound of the result of every interval operation has to be rounded downwards and the upper bound rounded upwards. Thus, the rounding mode has to be reset for every arithmetic operation. If setting the rounding mode and the arithmetic operation are equally fast this slows down interval arithmetic unnecessarily by a factor of two in comparison to conventional floating-point arithmetic. On all existing commercial processors, however, setting the rounding mode takes a multiple (three, ten, twenty and even more) of the time that is needed for the arithmetic operation. Thus an interval operation is unnecessarily at least eight (or twenty and even more) times slower than the corresponding floating-point operation. The rounding should be part of the arithmetic operation as required by the theory of computer arithmetic [33, 34]. Every one of the rounded operations $\boxdot, \triangledown, \triangle, \circ \in \{+, -, *, /\}$ with rounding to nearest, downwards or upwards should be equally fast and executed in a single cycle.

The IEEE arithmetic standard requires that these 12 operations for floating-point numbers give computed results that coincide with the rounded exact result of the operation for any operands [78]. The standard was developed around 1980 as a standard for microprocessors at a time when the typical microprocessor was the 8086 running at 2 MHz and serving a memory space of 64 KB. Since that time the speed of microprocessors has been increased by a factor of more than 1000. IEEE arithmetic is now even provided by supercomputers, the speed of which is still faster by magnitudes. Advances in computer technology are now so profound that the arithmetic capability and repertoire of computers can and should be expanded. In contrast to IEEE arithmetic a general theory of advanced computer arithmetic requires that all arithmetic operations in the usual product spaces of computation: the complex numbers, real and complex vectors, real and complex matrices, real and complex intervals as well as real and complex interval vectors and interval matrices are provided on the computer by a general mathematical mapping principle which is called a semimorphism. For definition see [33, 34]. This guarantees, among other things, that all arithmetic operations in all these spaces deliver a computed result which differs from the exact result of the operation by (no or) only a single rounding.

A careful analysis within the theory of computer arithmetic shows that the arithmetic operations in the computer representable subsets of these spaces can be realized on the computer by a modular technique provided fifteen fundamental operations are made available on a low level, possibly by fast hardware routines. These fifteen operations are

$$\boxplus, \;\boxminus, \;\boxtimes, \;\boxslash, \;\boxdot,$$
$$\triangledown, \;\triangledown, \;\triangledown, \;\triangledown, \;\triangledown,$$
$$\triangle, \;\triangle, \;\triangle, \;\triangle, \;\triangle.$$

Here $\boxdot, \circ \in \{+, -, *, /\}$ denotes operations using a monotone and antisymmetric rounding $\square$ from the real numbers onto the subset of floating-point numbers, such as rounding to the nearest floating-point number. Likewise $\triangledown$ and $\triangle$, $\circ \in \{+, -, *, /\}$ denote the operations using the monotone rounding downwards $\triangledown$ and upwards $\triangle$ respectively. $\boxdot, \triangledown$ and $\triangle$ denote scalar products with only a single rounding. That is, if $a = (a_i)$ and $b = (b_i)$ are vectors with floating-point components a_i, b_i, then $a \odot b := \bigcirc(a_1 * b_1 + a_2 * b_2 + \ldots + a_n * b_n)$, $\bigcirc \in \{\square, \triangledown, \triangle\}$. The multiplication and addition signs on the right hand side of the assignment denote exact multiplication and summation in the sense of real numbers.

Of these 15 fundamental operations above, traditional numerical methods use only the four operations $\boxplus, \boxminus, \boxtimes$ and $\boxslash$. Conventional interval arithmetic employs the eight operations $\triangledown, \triangledown, \triangledown, \triangledown$ and $\triangle, \triangle, \triangle, \triangle$. These eight operations are computer equivalents of the operations for real intervals; they provide interval arithmetic. The IEEE arithmetic standard requires 12 of these 15 fundamental operations: $\boxdot, \triangledown, \triangle$, $\circ \in \{+, -, *, /\}$. Generally speaking, interval arithmetic brings guarantees into computation, while the three scalar products $\boxdot, \triangledown$ and $\triangle$ bring high accuracy.

A detailed discussion of the implementation of the three scalar products on all kinds of computers is given in the first chapter. Basically the products $a_i * b_i$ are accumulated in fixed-point arithmetic with or without a single rounding at the very end of the accumulation. In contrast to accumulation in floating-point arithmetic, fixed-point accumulation is error free. Apart from this important property it is simpler than accumulation in floating-point and it is even faster. Accumulations in floating-point arithmetic are very sensitive with respect to cancellation.

So accumulations should be done in fixed-point arithmetic whenever possible whether the data are integers, floating-point numbers or products of two floating-point numbers. An arithmetic operation which can always be performed correctly on a digital computer should not be simulated by a routine which can easily fail in critical situations. Many real life and expensive accidents have been attributed to loss of numeric accuracy in a floating-point calculation or to other arithmetic failures. Examples are: bursting of a large turbine under test due to wrongly predicted eigenvalues; failure of early space shuttle retriever arms under space conditions; disastrous homing failure on ground to air missile missions; software failure in the Ariane 5 guidance program.

Advanced computer arithmetic requires a correct implementation of all arithmetic operations in the usual product spaces of computations. This includes interval arithmetic and in particular the three scalar products $\square, \triangledown$ and $\triangle$. This confronts us with another severe slowdown of interval arithmetic.

All commercial processors that provide IEEE arithmetic only deliver a rounded product to the outside world in the case of multiplication. Computation of an accurate scalar product requires products of the full double length. So these products have to be simulated on the processor. This slows down the multiplication by a factor of up to 10 in comparison to a rounded hardware multiplication. In a software simulation of the accurate scalar product the products of double length then have to be accumulated in fixed-point mode. This process is again slower by a factor of about 5 in comparison to a possibly wrong hardware accumulation of the products in floating-point arithmetic. Thus in summary a factor of at least 50 is the penalty for an accurate computation of the scalar product on existing processors. This is too much to be readily accepted by the user. In contrast to this a hardware implementation of the optimal scalar product could even be faster than a conventional implementation in floating-point arithmetic.

Another severe shortcoming which makes interval arithmetic slow is the fact that no reasonable interface to the programming languages has been accepted by the standardization committees so far. Operator overloading is not adequate for calling all fifteen operations $\square, \triangledown, \triangle, \circ \in \{+, -, *, /, \cdot\}$, in a high level programming language. A general operator concept is necessary for ease of programming (three real operations for $+, -, *, /$ and the dot product with three different roundings) otherwise clumsy and slow function calls have to be used to call different rounded arithmetic operations.

All these factors which make interval arithmetic on existing processors slow are quite well known. Nevertheless, they are generally not taken into account when the speed of interval methods is judged. It is, however, important that these factors are well understood. Real progress depends critically on an understanding of their details. Interval methods are not slow per se. It is the actual available arithmetic on existing processors which makes them slow. With better processor and language support, rigorous methods could be comparable in speed to today's "approximate" methods. Interval mathematics or mathematical numerics has been developed to a level where already today library routines could speedily deliver validated bounds instead of just approximations for small and medium size problems. This would ease the life of many users dramatically.

Future computers must be equipped with fast and effective interval arithmetic. At processor speeds of gigaFLOPS it is almost the only way to check the accuracy of a computation. Computer-generated graphics requires validation techniques in many cases.

After Sunaga's early paper the publication of Ramon E. Moore's book on interval arithmetic in 1966 [44] certainly was another milestone in the development of interval arithmetic. Moore's book is full of unconventional ideas which were out of the mainstream of numerical analysis of that time. To many colleagues the book appeared as an utopian dream. Others tried to carry out his ideas with little success in general. Computers were very very slow at that time. Today Moore's book appears as an exposition of extraordinary intellectual and creative power. The basic ideas of a great many well established methods of validation numerics can be traced back to Moore's book.

We conclude this introduction with a brief sketch of the development of interval arithmetic at the author's institute. Already by 1967 an ALGOL-60 extension implemented on a Zuse Z 23 computer provided operators and a number of elementary functions for a new data type *interval* [69, 70]. In 1968/69 this language was implemented on a more powerful computer, an Electrologica X8. To speed up the arithmetic, the hardware of the processor was extended by the four arithmetic operations with rounding downwards $\triangledown$, $\circ \in \{+, -, *, /\}$. Operations with rounding upwards were produced by use of the relation $\triangle(a) = -\triangledown(-a)$. Many early interval methods have been developed using these tools. Based on this experience a book [5] was written by two collaborators of that time. The English translation which appeared in 1983 is still a standard monograph on interval arithmetic [6].

At about 1969 the author became aware that interval and floating-point arithmetic basically follow the same mathematical mapping principles, and can be subsumed by a general mathematical theory of what is called advanced computer arithmetic in this paper. The basic assumption is that all arithmetic operations on computers (for real and complex numbers, real and complex intervals as well as for vectors and matrices over these four basic data types) should be defined by four simple rules which are called a semimorphism. This guarantees the best possible answers for all these arithmetic operations. A book on the subject was published in 1976 [33] and the German company

Nixdorf funded an implementation of the new arithmetic. At that time a Z-80 microprocessor with 64 KB main memory had to be used. The result was a PASCAL extension called PASCAL-SC, published in [37, 38]. The language provides about 600 predefined arithmetic operations for all the data types mentioned above and a number of elementary functions for the data types real and interval. The programming convenience of PASCAL-SC allowed a small group of collaborators to implement a large number of problem solving routines with automatic result verification within a few months. All this work was exhibited at the Hannover fair in March 1980 with the result that Nixdorf donated a number of computers to the Universität Karlsruhe. This allowed the programming education at Universität Karlsruhe to be decentralized from the summer of 1980. PASCAL-SC was the proof that advanced computer arithmetic need not be restricted to the very large computers. It had been realized on a microprocessor. When the PC appeared on the scene in 1982 it looked poor compared with what we had already two years earlier. But the PASCAL-SC system was never marketed.

In 1978 an English version of the theoretical foundation of advanced computer arithmetic was prepared during a sabbatical of the author jointly with W. L. Miranker at the IBM Research Center at Yorktown Heights. It appeared as a book in 1981 [34].

In May 1980 IBM became aware of the decentralized programming education with PASCAL-SC at the Universität Karlsruhe. This was the beginning of nearly ten years of close cooperation with IBM. We jointly developed and implemented a Fortran extension corresponding to the PASCAL extension with a large number of problem solving routines with automatic result verification [75–77].

In 1980 IBM had only the /370 architecture on the market. So we had to work for this architecture. IBM supported the arithmetic on an early processor (4361 in 1983) by microcode and later by VLSI design. Everything we developed for IBM was offered on the market as IBM program products in several versions between 1983 and 1989. But the products did not sell in the quantities IBM had expected. During the 1980s scientific computing had moved from the old mainframes to workstations and supercomputers. So the final outcome of these wonderful products was the same as for all the other earlier attempts to establish interval arithmetic effectively. With the next processor generation or a new language standard work for a particular processor loses its attraction and its value.

Nevertheless all these developments have contributed to the high standard attained by interval mathematics or mathematical numerics today. What we have today is a new version of PASCAL-SC, called PASCAL-XSC [26, 27, 29], with fast elementary functions and a corresponding C++ extension called C-XSC [28]. Both languages are translated into C so that they can be used on nearly all platforms. The arithmetic is implemented in software in C with all the regrettable consequences with respect to speed discussed earlier. Toolbox publications with problem solving routines are available for both languages [17, 18, 31].

Of course, much valuable work on the subject had been done at other places as well. International Conferences where new results can be presented and discussed are held regularly.

After completion of this paper Sun Microsystems announced an interval extension of Fortran 95 [83]. With this new product and compiler, interval arithmetic is now available on computers which are wide spread.

As Teruo Sunaga did in 1958 and many others after him, I am looking forward to, expect, and eagerly await a revision of the structure of the digital computer for better support of interval arithmetic.

3.2 Interval Arithmetic, a Powerful Calculus to Deal with Inequalities

Problems in technology and science are often described by an equation or a system of equations. Mathematics is used to manipulate these equations in order to obtain a solution. The Gauss algorithm, for instance, is used to compute the solution of a system of linear equations by adding, subtracting, multiplying and dividing equations in a systematic manner. Newton's method is used to compute approximately the location of a zero of a non linear function or of a system of such functions.

Data are often given by bounds rather than by simple numbers. Bounds are expressed by inequalities. To compute bounds for problems derived from given data requires a systematic calculus to deal with inequalities. Interval arithmetic provides this calculus. It supplies the basic rules for how to add, subtract, multiply, divide, and manipulate inequalities in a systematic manner: Let bounds for two real numbers a and b be given by the inequalities $a_1 \leq a \leq a_2$ and $b_1 \leq b \leq b_2$. Addition of these inequalities leads to bounds for the sum $a + b$:

$$a_1 + b_1 \leq a + b \leq a_2 + b_2.$$

The inequality for b can be reversed by multiplication with -1: $-b_2 \leq -b \leq -b_1$. Addition to the inequality for a then delivers the rule for the subtraction of two inequalities:

$$a_1 - b_2 \leq a - b \leq a_2 - b_1.$$

Interval arithmetic provides a shorthand notation for these rules suppressing the $\leq$ symbols. We simply identify the inequality $a_1 \leq a \leq a_2$ with the closed and bounded real interval $[a_1, a_2]$. The rules for addition and subtraction for two such intervals now read:

$$[a_1, a_2] + [b_1, b_2] = [a_1 + b_1, a_2 + b_2], \tag{3.1}$$

$$[a_1, a_2] - [b_1, b_2] = [a_1 - b_2, a_2 - b_1]. \tag{3.2}$$

The rule for multiplication of two intervals is more complicated. Nine cases are to be distinguished depending on whether a_1, a_2, b_1, b_2, are less or greater than zero. For division the situation is similar. Since we shall build upon these rules later they are cited here. For a detailed derivation see [33, 34]. In the tables the order relation $\leq$ is used for intervals. It is defined by

$$[a_1, a_2] \leq [b_1, b_2] :\iff a_1 \leq b_1 \wedge a_2 \leq b_2.$$

Table 3.1. The 9 cases for the multiplication of two intervals or inequalities

Nr.	$A = [a_1, a_2]$	$B = [b_1, b_2]$	$A * B$
1	$A \geq [0,0]$	$B \geq [0,0]$	$[a_1 b_1, a_2 b_2]$
2	$A \geq [0,0]$	$B \leq [0,0]$	$[a_2 b_1, a_1 b_2]$
3	$A \geq [0,0]$	$0 \in \overset{\circ}{B}$	$[a_2 b_1, a_2 b_2]$
4	$A \leq [0,0]$	$B \geq [0,0]$	$[a_1 b_2, a_2 b_1]$
5	$A \leq [0,0]$	$B \leq [0,0]$	$[a_2 b_2, a_1 b_1]$
6	$A \leq [0,0]$	$0 \in \overset{\circ}{B}$	$[a_1 b_2, a_1 b_1]$
7	$0 \in \overset{\circ}{A}$	$B \geq [0,0]$	$[a_1 b_2, a_2 b_2]$
8	$0 \in \overset{\circ}{A}$	$B \leq [0,0]$	$[a_2 b_1, a_1 b_1]$
9	$0 \in \overset{\circ}{A}$	$0 \in \overset{\circ}{B}$	$[\min(a_1 b_2, a_2 b_1), \max(a_1 b_1, a_2 b_2)]$

$$\tag{3.3}$$

Table 3.2. The 6 cases for the division of two intervals or inequalities

Nr.	$A = [a_1, a_2]$	$B = [b_1, b_2]$	A/B
1	$A \geq [0,0]$	$0 < b_1 \leq b_2$	$[a_1/b_2, a_2/b_1]$
2	$A \geq [0,0]$	$b_1 \leq b_2 < 0$	$[a_2/b_2, a_1/b_1]$
3	$A \leq [0,0]$	$0 < b_1 \leq b_2$	$[a_1/b_1, a_2/b_2]$
4	$A \leq [0,0]$	$b_1 \leq b_2 < 0$	$[a_2/b_1, a_1/b_2]$
5	$0 \in \overset{\circ}{A}$	$0 < b_1 \leq b_2$	$[a_1/b_1, a_2/b_1]$
6	$0 \in \overset{\circ}{A}$	$b_1 \leq b_2 < 0$	$[a_2/b_2, a_1/b_2]$

$$\tag{3.4}$$

In Tables 3.1 and 3.2 $\overset{\circ}{A}$ denotes the interior of A, i.e. $c \in \overset{\circ}{A}$ means $a_1 < c < a_2$. In the cases $0 \in B$ division A/B is not defined.

3.2 Interval Arithmetic, a Powerful Calculus to Deal with Inequalities

As a result of these rules it can be stated that in the case of real intervals the result of an interval operation $A \circ B$, for all $\circ \in \{+, -, *, /\}$, can be expressed in terms of the bounds of the interval operands (with the A/B exception above). In order to get each of these bounds, typically only one real operation is necessary. Only in case 9 of Table 3.1, $0 \in \mathring{A}$ and $0 \in \mathring{B}$, do two products have to be calculated and compared.

Whenever in the Tables 3.1 and 3.2 both operands are comparable with the interval $[0,0]$ with respect to $\leq, \geq, <$ or $>$, the result of the interval operation $A * B$ or A/B contains both bounds of A and B. If one or both of the operands A or B, however, contains zero as an interior point, then the result $A * B$ and A/B is expressed by only three of the four bounds of A and B. In all these cases (3, 6, 7, 8, 9) in Table 3.1, the bound which is missing in the expression for the result can be shifted towards zero without changing the result of the operation $A * B$. Similarly, in cases 5 and 6 in Table 3.2, the bound of B, which is missing in the expression for the resulting interval, can be shifted toward ∞ (resp. $-\infty$) without changing the result of the operation. This shows a certain lack of sensitivity of interval arithmetic or computing with inequalities whenever in the cases of multiplication and division one of the operands contains zero as an interior point.

In all these cases — 3, 6, 7, 8, 9, of Table 3.1 and 5, 6 of Table 3.2 — the result of $A * B$ or A/B also contains zero, and the formulas show that the result tends toward the zero interval if the operands that contain zero do likewise. In the limit case when the operand that contains zero has become the zero interval, no such imprecision is left. This suggests that within arithmetic expressions interval operands that contain zero as an interior point should be made as small in diameter as possible.

We illustrate the efficiency of this calculus for inequalities by a simple example. See [4]. Let $x = Ax + b$ be a system of linear equations in fixed point form with a contracting real matrix A and a real vector b, and let the interval vector X be a rough initial enclosure of the solution $x^* \in X$. We can now formally write down the Jacobi method, the Gauss-Seidel method, a relaxation method or some other iterative scheme for the solution of the linear system. In these formulas we then interpret all components of the vector x as being intervals. Doing so we obtain a number of iterative methods for the computation of enclosures of linear systems of equations. Further iterative schemes then can be obtained by taking the intersection of two successive approximations. If we now decompose all these methods in formulas for the bounds of the intervals we obtain a major number of methods for the computation of bounds for the solution of linear systems which have been derived by well-known mathematicians painstakingly about 40 years ago, see [14]. The calculus of interval arithmetic reproduces these and other methods in the simplest way. The user does not have to take care of the many case distinctions occurring in the matrix vector multiplications. The computer executes them automatically by the preprogrammed calculus. Also the rounding errors are enclosed. The calculus evolves its own dynamics.

3.3 Interval Arithmetic as Executable Set Operations

The rules (3.1), (3.2), (3.3), and (3.4) also can be interpreted as arithmetic operations for sets. As such they are special cases of general set operations. Further important properties of interval arithmetic can immediately be obtained via set operations. Let M be any set with a dyadic operation $\circ : M \times M \to M$ defined for its elements. The powerset $I\!PM$ of M is defined as the set of all subsets of M. The operation $\circ$ in M can be extended to the powerset $I\!PM$ by the following definition

$$A \circ B := \{a \circ b | a \in A \wedge b \in B\} \text{ for all } A, B \in I\!PM. \tag{3.5}$$

The least element in $I\!PM$ with respect to set inclusion as an order relation is the empty set. The greatest element is the set M. We denote the empty set by the character string [ ]. The empty set is subset of any set. Any arithmetic operation on the empty set produces the empty set.

The following properties are obvious and immediate consequences of (3.5):

$$A \subseteq B \wedge C \subseteq D \Rightarrow A \circ C \subseteq B \circ D \text{ for all } A, B, C, D \in I\!PM, \tag{3.6}$$

and in particular

$$a \in A \wedge b \in B \Rightarrow a \circ b \in A \circ B \text{ for all } A, B \in I\!PM. \tag{3.7}$$

(3.6) is called the *inclusion isotony* (or *inclusion monotony*). (3.7) is called the *inclusion property*.

By use of parentheses these rules can immediately be extended to expressions with more than one arithmetic operation, e.g.

$$A \subseteq B \wedge C \subseteq D \wedge E \subseteq F \Rightarrow A \circ C \subseteq B \circ D \Rightarrow (A \circ C) \circ E \subseteq (B \circ D) \circ F,$$

and so on. Moreover, if more than one operation is defined in M this chain of conclusions also remains valid for expressions containing several different operations.

If we now replace the general set M by the set of real numbers, (3.5), (3.6), and (3.7) hold in particular for the powerset $I\!P I\!R$ of the real numbers $I\!R$. This is the case for all operations $\circ \in \{+, -, *, /\}$, if we assume that in case of division 0 is not an element of the denominator, for instance, $0 \notin B$ in (3.5).

The set $I I\!R$ of closed and bounded intervals over $I\!R$ is a subset of $I\!P I\!R$. Thus (3.5), (3.6), and (3.7) are also valid for elements of $I I\!R$. The set $I I\!R$

with the operations (3.5), $\circ \in \{+, -, *, /\}$, is an algebraically closed[1] subset within $\mathbb{PR}$. That is, if (3.5) is performed for two intervals $A, B \in \mathbb{IR}$ the result is always an interval again. This holds for all operations $\circ \in \{+, -, *, /\}$ with $0 \notin B$ in case of division. This property is a simple consequence of the fact that for all arithmetic operations $\circ \in \{+, -, *, /\}$, $a \circ b$ is a continuous function of both variables. $A \circ B$ is the range of this function over the product set $A \times B$. Since A and B are closed intervals, $A \times B$ is a simply connected, bounded and closed subset of $\mathbb{R}^2$. In such a region the continuous function $a \circ b$ takes a maximum and a minimum as well as all values in between. Therefore

$$A \circ B = [\min_{a \in A, b \in B}(a \circ b), \max_{a \in A, b \in B}(a \circ b)], \text{ for all } \circ \in \{+, -, *, /\},$$

provided that $0 \notin B$ in case of division.

Consideration of (3.5), (3.6), and (3.7) for intervals of $\mathbb{IR}$ leads to the crucial properties of all applications of interval arithmetic. Because of the great importance of these properties we repeat them here explicitly. Thus we obtain for all operations $\circ \in \{+, -, *, /\}$:

The *set definition* of interval arithmetic:

$$A \circ B := \{a \circ b | a \in A \land b \in B\} \text{ for all } A, B \in \mathbb{IR}, \quad 0 \notin B \text{ in case of division,} \tag{3.8}$$

the *inclusion isotony* (or *inclusion monotony*):

$$A \subseteq B \land C \subseteq D \Rightarrow A \circ C \subseteq B \circ D \text{ for all } A, B, C, D \in \mathbb{IR}, \quad 0 \notin C, D \text{ in case of division,} \tag{3.9}$$

and in particular the *inclusion property*

$$a \in A \land b \in B \Rightarrow a \circ b \in A \circ B \text{ for all } A, B \in \mathbb{IR}, \quad 0 \notin B \text{ in case of division.} \tag{3.10}$$

If for $M = \mathbb{R}$ in (3.5) the number of elements in A or B is infinite, the operations are effectively not executable because infinitely many real operations would have to be performed. If A and B are intervals of $\mathbb{IR}$, however, the situation is different. In general A or B or both will again contain infinitely many real numbers. The result of the operation (3.8), however, can now be performed by a finite number of operations with real numbers, with the bounds of A and B. For all operations $\circ \in \{+, -, *, /\}$ the result is obtained by the explicit formulas (3.1), (3.2), (3.3), and (3.4), [33, 34].

For intervals $A = [a_1, a_2]$ and $B = [b_1, b_2]$ the formulas (3.1), (3.2), (3.3), and (3.4) can be summarized by

$$A \circ B = [\min_{i,j=1,2}(a_i \circ b_j), \max_{i,j=1,2}(a_i \circ b_j)] \text{ for all } \circ \in \{+, -, *, /\}, \tag{3.11}$$

---

[1] as the integers are within the reals for $\circ \in \{+, -, *\}$

94 3. Interval Arithmetic Revisited

with $0 \notin B$ in case of division.

Since interval operations are particular powerset operations, the inclusion isotony and the inclusion property also hold for expressions with more than one arithmetic operation.

In programming languages the concept of an arithmetic expression is usually defined to be a little more general. Besides constants and variables elementary functions (sometimes called standard functions) like sqr, sqrt, sin, cos, exp, ln, tan, ... may also be elementary ingredients. All these are put together with arithmetic operators and parentheses into the general concept of an arithmetic expression. This construct is illustrated by the syntax diagram of Fig. 3.1. Therein solid lines are to be traversed from left to right and from top to bottom. Dotted lines are to be traversed oppositely, i.e. from right to left and from bottom to top. In Fig. 3.1 the syntax variable REAL FUNCTION merely represents a real arithmetic expression hidden in a subroutine.

Now we define the general concept of an arithmetic expression for the new data type interval by exchanging the data type *real* in Fig. 3.1 for the new data type *interval*. This results in the syntax diagram for INTERVAL EXPRESSION shown in Fig. 3.2. In Fig. 3.2 the syntax variable INTERVAL FUNCTION represents an interval expression hidden in a subroutine.

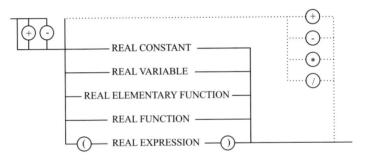

Fig. 3.1. Syntax diagram for REAL EXPRESSION

In the syntax diagram for INTERVAL EXPRESSION in Fig. 3.2 the concept of an interval elementary function is not yet defined. We simply define it as the range of function values taken over an interval (out of the domain of definition $D(f)$ of the function). In case of a real function f we denote the range of values over the interval $[a_1, a_2]$ by

$$f([a_1, a_2]) := \{f(a) \mid a \in [a_1, a_2]\} \text{ with } [a_1, a_2] \in D(f).$$

For instance:

3.3 Interval Arithmetic as Executable Set Operations

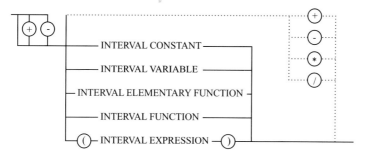

Fig. 3.2. Syntax diagram for INTERVAL EXPRESSION

$$e^{[a_1, a_2]} = [e^{a_1}, e^{a_2}],$$

$$[a_1, a_2]^{2n} = \begin{cases} [\min(a_1^{2n}, a_2^{2n}), \max(a_1^{2n}, a_2^{2n})] & \text{for } 0 \notin [a_1, a_2], \\ [0, \max(a_1^{2n}, a_2^{2n})] & \text{for } 0 \in [a_1, a_2], \end{cases}$$

$$\sin[-\frac{\pi}{4}, \frac{\pi}{4}] = [-\frac{1}{2}\sqrt{2}, \frac{1}{2}\sqrt{2}],$$

$$\cos[0, \frac{\pi}{4}] = [0, 1].$$

For non monotonic functions the computation of the range of values over an interval $[a_1, a_2]$ requires the determination of the global minimum and maximum of the function in the interval $[a_1, a_2]$. For the usual elementary functions, however, these are known. With this definition of elementary functions for intervals the key properties of interval arithmetic, the inclusion monotony (3.7) and the inclusion property (3.8) extend immediately to elementary functions and with this to interval expressions as defined in Fig. 3.2:

$$A \subseteq B \Rightarrow f(A) \subseteq f(B), \text{ with } A, B \in I\!R \qquad \text{inclusion isotone,}$$

and in particular for $a \in \mathbb{R}$ and $A \in I\!R$:

$$a \in A \Rightarrow f(a) \in f(A) \qquad \text{inclusion property.}$$

We summarize the development so far by stating that interval arithmetic expressions are generally inclusion isotone and that the inclusion property holds. These are the key properties of interval arithmetic. They give interval arithmetic its raison d'être. To start with, they provide the possibility of enclosing imprecise data within bounds and then continuing the computation with these bounds. This always results in guaranteed enclosures.

As the next step we define a (computable) real function simply by a real arithmetic expression. We need the concept of an *interval evaluation of a real function*. It is defined as follows: In the arithmetic expression for the function all operands are replaced by intervals and all operations by interval operations (where all intervals must be within the domain of definition of the real operands). This is just the step from Fig. 3.1 to Fig. 3.2. What is obtained is an interval expression. Then all arithmetic operations are performed in interval arithmetic. For a real function $f(a)$ we denote the interval evaluation over the interval A by $F(A)$.

With this definition we can immediately conclude that interval evaluations of (computable) real functions are inclusion isotone and that the inclusion property holds in particular:

$$A \subseteq B \Rightarrow F(A) \subseteq F(B) \qquad \text{inclusion isotone,} \qquad (3.12)$$

$$a \in A \Rightarrow f(a) \in F(A) \qquad \text{inclusion property.} \qquad (3.13)$$

These concepts immediately extend in a natural way to functions of several real variables. In this case in (3.13) a is an n-tuple, $a = (a_1, a_2, \ldots, a_n)$, and A and B are higher dimensional intervals, e.g. $A = (A_1, A_2, \ldots, A_n)$.

Remark: Two different real arithmetic expressions can define equivalent real functions, for instance:

$$f(x) = x(x-1) \quad \text{and} \quad g(x) = x^2 - x.$$

Evaluation of the two expressions for a real number always leads to the same real function value. In contrast to this, interval evaluation of the two expressions may lead to different intervals. In the example we obtain for the interval $A = [1, 2]$:

$$F(A) = [1,2]([1,2] + [-1,-1]) \qquad G(A) = [1,2][1,2] - [1,2]$$
$$= [1,2][0,1] = [0,2], \qquad\qquad = [1,4] - [1,2] = [-1,3].$$

Although an interval evaluation of a real function is very naturally defined via the arithmetic expression of the function, a closer look at the syntax diagram in Fig. 3.2 reveals major problems that appear when such evaluations are to be coded. The widely used programming languages do not provide the necessary ease of programming. An interval evaluation of a real function should be performable as easily as an execution of the corresponding expression in real arithmetic. For that purpose the programming language

1. must allow an operator notation $A \circ B$ for the basic interval operations $\circ \in \{+, -, *, /\}$, i.e. operator overloading must be provided,
2. the concept of a function subroutine must not be restricted to the data types *integer* and *real*, i.e. subroutine functions with general result type should be provided by the programming language, and
3. the elementary functions must be provided for interval arguments.

While 1. and 2. are challenges for the designer of the programming language, 3. is a challenge for the mathematician. In a conventional call of an elementary function the computer provides a result, the accuracy of which cannot easily be judged by the user. This is no longer the case when the elementary functions are provided for interval arguments. Then, if called for a point interval (where the lower and upper bound coincide), a comparison of the lower and upper bound of the result of the interval evaluation of the function reveals immediately the accuracy with which the elementary function has been implemented. This situation has forced extremely careful implementation of the elementary functions and since interval versions of the elementary functions have been provided on a large scale [26–29, 37, 38, 77] the conventional real elementary functions on computers also had to be and have been improved step by step by the manufacturers. A most advanced programming environment in this respect is a decimal version of PASCAL-XSC [10] where, besides the usual 24 elementary functions, about the same number of special functions are provided for real and interval arguments with highest accuracy.

1., 2. and 3. are minimum requirements for any sophisticated use of interval arithmetic. If they are not met, coding difficulties absorb all the attention and capacity of users and prevent them from developing deeper mathematical ideas and insight. So far none of the widespread programming languages like Fortran, C, and even Fortran 95 and C++ provide the necessary programming ease. This is the basic reason for the slow progress in the field. It is a matter of fact that a great deal of the existing and established interval methods and algorithms have originally been developed in PASCAL-XSC even if they have been coded afterwards in other languages. Programming ease is essential indeed. The typical user, however, is reluctant to leave the programming environment he is used to, just to apply interval methods.

We summarize this discussion by stating that it does not suffice for an adequate use of interval arithmetic on computers that only the four basic arithmetic operations $+, -, *$ and $/$ for intervals are somehow supported by the computer hardware. An appropriate language support is absolutely necessary. So far this has been missing. This is the basic dilemma of interval arithmetic. Experience has shown that it cannot be overcome via slow moving standardization committees for programming languages. Two things seem to be necessary for the great breakthrough. A major vendor has to provide the necessary support and the body of numerical analysts must acquire a broader insight and skills in order to use this support.

3.4 Enclosing the Range of Function Values

The interval evaluation of a real function f over the interval A was denoted by $F(A)$. We now compare it with the range of function values over the interval A which was denoted by

$$f(A) := \{f(a) \mid a \in A\}. \tag{3.14}$$

We have observed that interval evaluation of an arithmetic expression and of real functions is inclusion isotone (3.9), (3.12) and that the inclusion property (3.10), (3.13) holds. Since (3.10) and (3.13) hold for all $a \in A$ we can immediately state that

$$f(A) \subseteq F(A), \qquad (3.15)$$

i.e. that the interval evaluation of a real function over an interval delivers a superset of the range of function values over that interval. If A is a point interval $[a, a]$ this reduces to:

$$f(a) \in F([a, a]). \qquad (3.16)$$

These are basic properties of interval arithmetic. Computing with inequalities always aims for bounds for function values, or for bounds for the range of function values. Interval arithmetic allows this computation in principle.

The range of function values over an interval is needed for many applications. Its computation is a very difficult task. It is equivalent to the computation of the global minimum and maximum of the function in that interval. An interval evaluation of the arithmetic expression on the other hand is very easy to perform. It requires about twice as many real arithmetic operations as an evaluation of the function in real arithmetic. Thus interval arithmetic provides an easy means to compute upper and lower bounds for the range of function values.

In the end a complicated algorithm just performs an arithmetic expression. So an interval evaluation of the algorithm would compute bounds for the result from given bounds for the data. However, it is observed that in doing so, in general, the diameters of the intervals grow very fast and for large algorithms the bounds quickly become meaningless in particular if the bounds for the data are already large. This raises the question whether measures can be taken to keep the diameters of the intervals from growing too fast. Interval arithmetic has developed such measures and we are going to sketch these now.

If an enclosure for a function value is computed by (3.16), the quality of the computed result $F([a, a])$ can be judged by the diameter of the interval $F([a, a])$. This possibility of easily judging the quality of the computed result, is not available in (3.15). Even if $F(A)$ is a large interval, it can be a good approximation for the range of function values $f(A)$ if the latter is large also. So some means to measure the deviation between $f(A)$ and $F(A)$ in (3.15) is desirable.

It is well known that the set $I\!R$ of real intervals becomes a metric space with the so called Hausdorff metric, where the distance q of two intervals $A = [a_1, a_2]$ and $B = [b_1, b_2]$ is defined by

$$q(A, B) := \max\{|a_1 - b_1|, |a_2 - b_2|\}. \qquad (3.17)$$

See, for instance, [6].

With this distance function q the following relation can be proved to hold under natural assumptions on f:

3.4 Enclosing the Range of Function Values

$$q(f(A), F(A)) \leq \alpha \cdot d(A), \text{ with a constant } \alpha \geq 0. \tag{3.18}$$

Here $d(A)$ denotes the diameter of the interval A:

$$d(A) := |a_2 - a_1|. \tag{3.19}$$

In case of functions of several real variables the maximum of the diameters $d(A_i)$ appears on the right hand side of (3.18).

The relation (3.18) shows that the distance between the range of values of the function f over the interval A and the interval evaluation of the expression for f tends to zero linearly with the diameter of the interval A. So the overestimation of $f(A)$ by $F(A)$ decreases with the diameter of A and in the limit $d(A) = 0$ no such overestimation is left.

Because of this result subdivision of the interval A into subintervals A_i, $i = 1(1)n$, with $A = \bigcup_{i=1}^{n} A_i$ is a frequently applied technique to obtain better approximations for the range of function values. Then (3.18) holds for each subinterval:

$$q(f(A_i), F(A_i)) \leq \alpha_i \cdot d(A_i), \text{ with } \alpha_i \geq 0 \text{ and } i = 1(1)n,$$

and, in general, the union of the interval evaluations over all subintervals

$$\bigcup_{i=1}^{n} F(A_i)$$

is a much better approximation for the range $f(A)$ than is $F(A)$.

There are yet other methods to obtain better enclosures for the range of function values $f(A)$. We have already observed that the interval evaluation $F(A)$ of a function f depends on the expression used for the representation of f. So by choosing appropriate representations for f the overestimation of $f(A)$ by the interval evaluation $F(A)$ can often be reduced. Indeed, if f allows a representation of the form

$$f(x) = f(c) + (x - c) \cdot h(x), \text{ with } c \in A, \tag{3.20}$$

then under natural assumptions on h the following inequality holds

$$q(f(A), F(A)) \leq \beta \cdot (d(A))^2, \text{ with a constant } \beta \geq 0. \tag{3.21}$$

(3.20) is called a centered form of f. In (3.20) c is not necessarily the center of A although it is often chosen as the center. (3.21) shows that the distance between the range of values of the function f over the interval A and the interval evaluation of a centered form of f tends toward zero quadratically with the diameter of the interval A. In practice, this means that for small intervals the interval evaluation of the centered form leads to a very good approximation of the range of function values over an interval A. Again, subdivision is a method that can be applied in the case of a large interval

A. It should be clear, however, that in general only for small intervals is the bound in (3.21) better than in (3.18).

The decrease of the overestimation of the range of function values by the interval evaluation of the function with the diameter of the interval A, and the method of subdivision, are reasons why interval arithmetic can successfully be used in many applications. Numerical methods often proceed in small steps. This is the case, for instance, with numerical quadrature or cubature, or with numerical integration of ordinary differential equations. In all these cases an interval evaluation of the remainder term of the integration formula (using differentiation arithmetic) controls the step size of the integration, and anyhow because of the small steps, overestimation is practically negligible.

We now mention briefly how centered forms can be obtained. Usually a centered form is derived via the mean-value theorem. If f is differentiable in its domain D, then $f(x) = f(c) + f'(\xi)(x - c)$ for fixed $c \in D$ and some ξ between x and c. If x and c are elements out of the interval $A \subseteq D$, then also $\xi \in A$. Therefore

$$f(x) \in F(A) := f(c) + F'(A)(A - c), \text{ for all } x \in A.$$

Here $F'(A)$ is an interval evaluation of $f'(x)$ in A.

In (3.20) the slope

$$h(x) = \frac{f(x) - f(c)}{x - c}$$

can be used instead of the derivative for the representation of $f(x)$. Slopes often lead to better enclosures for $f(A)$ than do derivatives. For details see [7, 32, 53].

Derivatives and enclosures of derivatives can be computed by a process which is called automatic differentiation or differentiation arithmetic. Slopes and enclosures of slopes can be computed by another process which is very similar to automatic differentiation. In both cases the computation of the derivative or slope or enclosures of these is done together with the computation of the function value. For these processes only the expression or algorithm for the function is required. No explicit formulas for the derivative or slope are needed. The computer interprets the arithmetic operations in the expression by differentiation or slope arithmetic. The arithmetic is hidden in the runtime system of the compiler. It is activated by type specification of the operands. For details see [17, 18, 53], and Section 3.8. Thus the computer is able to produce and enclose the centered form via the derivative or slope automatically.

Without going into further details we mention once more, that all these considerations are not restricted to functions of a single real variable. Subdivision in higher dimensions, however, is a difficult task which requires additional tools and strategies. Typical of such problems are the computation of the bounds of the solution of a system of nonlinear equations, and global optimization or numerical integration of functions of more than one real variable.

In all these and other cases, zero finding is a central task. Here the extended Interval Newton Method plays an extraordinary role so we are now going to review this method, which is also one of the requirements that have to be met when interval arithmetic is implemented on the computer.

3.5 The Interval Newton Method

Traditionally Newton's method is used to compute an approximation of a zero of a nonlinear real function $f(x)$, i.e. to compute a solution of the equation

$$f(x) = 0. \tag{3.22}$$

The method approximates the function $f(x)$ in the neighborhood of an initial value x_0 by the linear function (the tangent)

$$t(x) = f(x_0) + f'(x_0)(x - x_0) \tag{3.23}$$

the zero of which can easily be calculated by

$$x_1 := x_0 - \frac{f(x_0)}{f'(x_0)}. \tag{3.24}$$

x_1 is used as new approximation for the zero of (3.22). Continuation of this method leads to the general iteration scheme:

$$x_{\nu+1} := x_\nu - \frac{f(x_\nu)}{f'(x_\nu)}, \quad \nu = 0, 1, 2, \ldots. \tag{3.25}$$

It is well known that if $f(x)$ has a single zero x^* in an interval X and $f(x)$ is twice continuously differentiable, then the sequence

$$x_0, x_1, x_2, \ldots, x_\nu, \ldots$$

converges quadratically towards x^* if x_0 is sufficiently close to x^*. If the latter condition does not hold the method may well fail.

The interval version of Newton's method computes an enclosure of the zero x^* of a continuously differentiable function $f(x)$ in the interval X by the following iteration scheme:

$$X_{\nu+1} := (m(X_\nu) - \frac{f(m(X_\nu))}{F'(X_\nu)}) \cap X_\nu, \quad \nu = 0, 1, 2, \ldots, \tag{3.26}$$

with $X_0 = X$. Here $F'(X_\nu)$ is the interval evaluation of the first derivative $f'(x)$ of f over the interval X_ν and $m(X_\nu)$ is the midpoint of the interval X_ν. Instead of $m(X_\nu)$ another point within X_ν could be chosen. The method can only be applied if $0 \notin F'(X_0)$. This guarantees that $f(x)$ has only a single zero in X_0.

3. Interval Arithmetic Revisited

In contrast to (3.25) the method (3.26) can never diverge (fail). Because of the intersection with X_ν the sequence

$$X_0 \supseteq X_1 \supseteq X_2 \supseteq \ldots \tag{3.27}$$

is bounded. It can be shown that under natural conditions on the function f the sequence converges quadratically to x^* [6, 47].

The operator

$$N(X) := x - \frac{f(x)}{F'(X)}, \quad x \in X \in I\!\!R \tag{3.28}$$

is called the Interval Newton Operator. It has the following properties:

I. If $N(X) \subseteq X$, then $f(x)$ has exactly one zero x^* in X.
II. If $N(X) \cap X = [\,]$ then $f(x)$ has no zero in X.

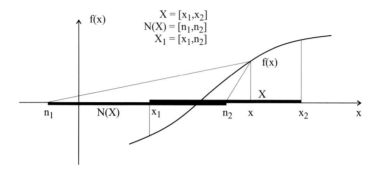

Fig. 3.3. Geometric interpretation of the Interval Newton Method

Thus, $N(X)$ can be used to prove the existence or absence of a zero x^* of $f(x)$ in X. Since in the case of existence of a zero x^* in X the sequence (3.26), (3.27) converges, in the case of absence the situation $N(X) \cap X = [\,]$ must occur in (3.27).

The interval version of Newton's method (3.26) can also be derived via the mean value theorem. If $f(x)$ is continuously differentiable and has a single zero x^* in the interval X, and $f'(x) \neq 0$ for all $x \in X$, then

$$f(x) = f(x^*) + f'(\xi)(x - x^*) \text{ for all } x \in X \text{ and some } \xi \text{ between } x \text{ and } x^*$$

Since $f(x^*) = 0$ and $f'(\xi) \neq 0$ this leads to

$$x^* = x - \frac{f(x)}{f'(\xi)}.$$

If $F'(X)$ denotes the interval evaluation of $f'(x)$ over the interval X, we have $f'(\xi) \in F'(X)$ and therefore

$$x^* = x - \frac{f(x)}{f'(\xi)} \in x - \frac{f(x)}{F'(X)} = N(X) \text{ for all } x \in X,$$

i.e. $x^* \in N(X)$ and thus

$$x^* \in (x - \frac{f(x)}{F'(X)}) \cap X = N(X) \cap X.$$

Now we obtain by setting $X_0 := X$ and $x = m(X_0)$

$$X_1 := (m(X_0) - \frac{f(m(X_0))}{F'(X_0)}) \cap X_0,$$

and by continuation (3.26).

In close similarity to the conventional Newton method the Interval Newton Method also allows some geometric interpretation. For that purpose let be $X = [x_1, x_2]$ and $N(X) = [n_1, n_2]$. $F'(X)$ is the interval evaluation of $f'(x)$ over the interval X. As such it is a superset of all slopes of tangents that can occur in X. (3.24) computes the zero of the tangent of $f(x)$ in $(x_0, f(x_0))$. Similarly $N(X)$ is the interval of all zeros of straight lines through $(x, f(x))$ with slopes within $F'(X)$, see Fig. 3.3. Of course, $f'(x) \in F'(X)$.

The straight line through $f(x)$ with the least slope within $F'(X)$ cuts the real axis in n_1, and the one with the greatest slope in n_2. Thus the Interval Newton Operator $N(X)$ computes the interval $[n_1, n_2]$ which in the sketch of Fig. 3.3 is situated on the left hand side of x. The intersection of $N(X)$ with X then delivers the new interval X_1. In the example in Fig. 3.3, $X_1 = [x_1, n_2]$.

Newton's method allows some visual interpretation. From the point $(x, f(x))$ the conventional Newton method sends a beam along the tangent. The search is continued at the intersection of this beam with the x-axis. The Interval Newton Method sends a set of beams like a floodlight from the point $(x, f(x))$ to the x-axis. This set includes the directions of all tangents that occur in the entire interval X. The interval $N(X)$ comprises all cuts of these beams with the x-axis.

It is a fascinating discovery that the Interval Newton Method can be extended so that it can be used to compute all zeros of a real function in a given interval. The basic idea of this extension is already old [3]. Many scientists have worked on details of how to use this method, of how to define the necessary arithmetic operations, and of how to bring them to the computer. But inconsistencies have occurred again and again. However, it seems that understanding has now reached a point which allows a consistent realization of the method and of the necessary arithmetic. The extended Interval Newton Method is the most powerful and most frequently used tool for subdivision in higher dimensional spaces. It requires an extension of interval arithmetic which we are now going to discuss.

3.6 Extended Interval Arithmetic

In the definition of interval arithmetic, division by an interval which includes zero was excluded. We are now going to eliminate this exception.

The real numbers $I\!R$ are defined as a conditionally complete, linearly ordered field.[2] With respect to the order relation $\leq$ they can be completed by adjoining a least element $-\infty$ and a greatest element $+\infty$. We denote the resulting set by $I\!R^* := I\!R \cup \{-\infty\} \cup \{+\infty\}$. $\{I\!R^*, \leq\}$ is a complete lattice.[3] This completion is frequently applied in mathematics and it is well known that the new elements $-\infty$ and $+\infty$ fail to satisfy several of the algebraic properties of a field. $-\infty$ and $+\infty$ are not real numbers! For example $a + \infty = b + \infty$ even if $a < b$, so that the cancellation law is not valid. For the new elements $-\infty$ and $+\infty$ the following operations with elements $x \in I\!R$ are defined in analysis:

$$\begin{aligned}
\infty + x &= \infty, & \infty * x &= \infty \text{ for } x > 0, \\
-\infty + x &= -\infty, & \infty * x &= -\infty \text{ for } x < 0, \\
\tfrac{x}{\infty} &= \tfrac{x}{-\infty} = 0, & & \\
\infty + \infty &= \infty * \infty = \infty, & & \\
-\infty + (-\infty) &= (-\infty) * \infty = -\infty.
\end{aligned} \qquad (3.29)$$

together with variants obtained by applying the sign rules and the law of commutativity. Not defined are the terms $\infty - \infty$ and $0 * \infty$, again with variants obtained by applying the sign rules and the law of commutativity. These rules are well established in real analysis and there is no need to extend them for the purposes of interval arithmetic in $I\!I\!R$.

$I\!R$ is a set with certain arithmetic operations. These operations can be extended to the powerset $I\!P I\!R$ in complete analogy to (3.5):

$$A \circ B := \{a \circ b \mid a \in A \wedge b \in B\} \text{ for all } \circ \in \{+, -, *, /\}, \text{ and all } A, B \in I\!P I\!R. \qquad (3.30)$$

As a consequence of (3.30) again the *inclusion isotony* (3.6) and the *inclusion property* (3.7) hold for all operations and arithmetic expressions in $I\!P I\!R$. In particular, this is the case if (3.30) is restricted to operands of $I\!I\!R$. $I\!I\!R$ is a subset of $I\!P I\!R$.

We are now going to define division by an interval B of $I\!I\!R$ which contains zero. It turns out that the result is no longer an interval of $I\!I\!R$. But we can apply the definition of the division in the powerset as given by (3.30). This leads to

$$A/B := \{a/b \mid a \in A \wedge b \in B\} \text{ for all } A, B \in I\!I\!R. \qquad (3.31)$$

---

[2] An ordered set is called conditionally complete if every non empty, bounded subset has a greatest lower bound (infimum) and a least upper bound (supremum).
[3] In a complete lattice every subset has an infimum and a supremum.

3.6 Extended Interval Arithmetic

In order to interpret the right hand side of (3.31) we remember that the quotient a/b is defined as the inverse operation of multiplication, i.e. as the solution of the equation $b \cdot x = a$. Thus (3.31) can also be written in the form

$$A/B := \{x \mid bx = a \wedge a \in A \wedge b \in B\} \quad \text{for all } A, B \in I\!R. \tag{3.32}$$

Now we have to interpret the right hand side of (3.32). We are interested in obtaining simply executable, explicit formulas for the right hand side of (3.32). The case $0 \notin B$ was already dealt with in Table 3.2. So we assume here generally that $0 \in B$. For $A = [a_1, a_2]$ and $B = [b_1, b_2] \in I\!R$, $0 \in B$ the following eight distinct cases can be set out:

1. $0 \in A$, $0 \in B$.
2. $0 \notin A$, $B = [0, 0]$.
3. $a_1 \leq a_2 < 0$, $b_1 < b_2 = 0$.
4. $a_1 \leq a_2 < 0$, $b_1 < 0 < b_2$.
5. $a_1 \leq a_2 < 0$, $0 = b_1 < b_2$.
6. $0 < a_1 \leq a_2$, $b_1 < b_2 = 0$.
7. $0 < a_1 \leq a_2$, $b_1 < 0 < b_2$.
8. $0 < a_1 \leq a_2$, $0 = b_1 < b_2$.

The list distinguishes the cases $0 \in A$ (case 1) and $0 \notin A$ (cases 2 to 8). Since it is generally assumed that $0 \in B$ these eight cases indeed cover all possibilities.

We are now going to derive simple formulas for the result of the interval division A/B for these eight cases:

1. $0 \in A \wedge 0 \in B$. Since every $x \in I\!R$ fulfils the equation $0 \cdot x = 0$, we have $A/B = (-\infty, +\infty)$. Here $(-\infty, +\infty)$ denotes the open interval between $-\infty$ and $+\infty$ which just consists of all real numbers $I\!R$, i.e. $A/B = I\!R$.
2. In case $0 \notin A \wedge B = [0, 0]$ the set defined by (3.32) consists of all elements which fulfil the equation $0 \cdot x = a$ for $a \in A$. Since $0 \notin A$, there is no real number which fulfils this equation. Thus A/B is the empty set $A/B = [\,]$.

In all other cases $0 \notin A$ also. We have already observed under 2. that in this case the element 0 in B does not contribute to the solution set. So it can be excluded without changing the set A/B.

So the general rule for computing A/B by (3.32) is to punch out zero of the interval B and replace it by a small positive or negative number ϵ as the case may be. The so changed interval B is denoted by $\overline{B}$ and represented in column 4 of Table 3.3. With this $\overline{B}$ the solution set $A/\overline{B}$ can now easily be computed by applying the rules of Table 3.2. The results are shown in column 5 of Table 3.3. Now the desired result A/B as defined by (3.32) is obtained if in column 5 ϵ tends to zero. Thus in cases 3 to 8 the results are obtained by the limit process $A/B = \lim_{\epsilon \to 0} A/\overline{B}$. The solution set A/B is shown in the last column of Table 3.3 for all the 8 cases. There, as usual in mathematics

106   3. Interval Arithmetic Revisited

parentheses denote open interval ends, i.e. the bound is excluded. In contrast to this brackets denote closed interval ends, i.e. the bound is included.

In Table 3.3 the operands A and B of the division A/B are intervals of $I\!R$! The results of the division A/B shown in the last column, however, are no longer intervals of $I\!R$ nor are they intervals of $I\!R^*$ which is the set of intervals over R^*. This is logically correct and should not be surprising, since the division has been defined as an operation in $I\!P\!R$ by (3.30).

Table 3.4 shows the result of the division A/B of two intervals $A = [a_1, a_2]$ and $B = [b_1, b_2]$ in the case $0 \in B$ in a more convenient layout.

Table 3.3. The 8 cases of the division of two intervals A/B, with $A, B \in I\!R$ and $0 \in B$.

case	$A = [a_1, a_2]$	$B = [b_1, b_2]$	$\overline{B}$	$A/\overline{B}$	A/B
1	$0 \in A$	$0 \in B$			$(-\infty, +\infty)$
2	$0 \notin A$	$B = [0, 0]$			$[\ ]$
3	$a_2 < 0$	$b_1 < b_2 = 0$	$[b_1, -\epsilon]$	$[a_2/b_1, a_1/(-\epsilon)]$	$[a_2/b_1, +\infty)$
4	$a_2 < 0$	$b_1 < 0 < b_2$	$[b_1, -\epsilon] \cup [\epsilon, b_2]$	$[a_2/b_1, a1/(-\epsilon)] \cup [a_1/\epsilon, a_2/b_2]$	$(-\infty, a_2/b_2] \cup [a_2/b_1, +\infty)$
5	$a_2 < 0$	$0 = b_1 < b_2$	$[\epsilon, b_2]$	$[a_1/\epsilon, a_2/b_2]$	$(-\infty, a_2/b_2]$
6	$a_1 > 0$	$b_1 < b_2 = 0$	$[b_1, -\epsilon]$	$[a_2/(-\epsilon), a_1/b_1]$	$(-\infty, a_1/b_1]$
7	$a_1 > 0$	$b_1 < 0 < b_2$	$[b_1, -\epsilon] \cup [\epsilon, b_2]$	$[a_2/(-\epsilon), a1/b_1] \cup [a_1/b_2, a_2/\epsilon]$	$(-\infty, a_1/b_1] \cup [a_1/b_2, +\infty)$
8	$a_1 > 0$	$0 = b_1 < b_2$	$[\epsilon, b_2]$	$[a_1/b_2, a_2/\epsilon]$	$[a_1/b_2, +\infty)$

Table 3.4. The result of the division A/B, with $A, B \in I\!R$ and $0 \in B$.

A/B	$B = [0, 0]$	$b_1 < b_2 = 0$	$b_1 < 0 < b_2$	$0 = b_1 < b_2$
$a_2 < 0$	$[\ ]$	$[a_2/b_1, +\infty)$	$(-\infty, a_2/b_2] \cup [a_2/b_1, +\infty)$	$(-\infty, a_2/b_2]$
$a_1 \leq 0 \leq a_2$	$(-\infty, +\infty)$	$(-\infty, +\infty)$	$(-\infty, +\infty)$	$(-\infty, +\infty)$
$a_1 > 0$	$[\ ]$	$(-\infty, a_1/b_1]$	$(-\infty, a_1/b_1] \cup [a_1/b_2, +\infty)$	$[a_1/b_2, +\infty)$

For completeness we repeat at the end of this section the results of the basic arithmetic operations for intervals $A = [a_1, a_2]$ and $B = [b_1, b_2]$ of $I\!R$ which have already been given in Section 3.2. In the cases of multiplication and division we use different representations. We also list the basic rules of the order relations $\leq$ and $\subseteq$ for intervals of $I\!R^*$.

I. Equality: $[a_1, a_2] = [b_1, b_2] :\Leftrightarrow a_1 = b_1 \wedge a_2 = b_2$.

II. Addition: $[a_1, a_2] + [b_1, b_2] = [a_1 + b_1, a_2 + b_2]$.
III. Subtraction: $[a_1, a_2] - [b_1, b_2] = [a_1 - b_2, a_2 - b_1]$.
IV. Negation: $A = [a_1, a_2]$, $-A = [-a_2, -a_1]$.
V. Multiplication:

$A \cdot B$	$b_1 \geq 0$	$b_1 < 0 < b_2$	$b_2 \leq 0$
$a_2 \leq 0$	$[a_1 b_2, a_2 b_1]$	$[a_1 b_2, a_1 b_1]$	$[a_2 b_2, a_1 b_1]$
$a_1 < 0 < a_2$	$[a_1 b_2, a_2 b_2]$	$[\min(a_1 b_2, a_2 b_1),$ $\max(a_1 b_1, a_2 b_2)]$	$[a_2 b_1, a_1 b_1]$
$a_1 \geq 0$	$[a_1 b_1, a_2 b_2]$	$[a_2 b_1, a_2 b_2]$	$[a_2 b_1, a_1 b_2]$

VI. Division, $0 \notin B$:

A/B	$b_1 > 0$	$b_2 < 0$
$a_2 \leq 0$	$[a_1/b_1, a_2/b_2]$	$[a_2/b_1, a_1/b_2]$
$a_1 < 0 < a_2$	$[a_1/b_1, a_2/b_1]$	$[a_2/b_2, a_1/b_2]$
$a_1 \geq 0$	$[a_1/b_2, a_2/b_1]$	$[a_2/b_2, a_1/b_1]$

The closed intervals over the real numbers $I\!R^*$ are ordered with respect to two different order relations, the comparison $\leq$ and the set inclusion $\subseteq$. With respect to both order relations they are complete lattices. The basic properties are:

VII. $\{I\!R^*, \leq\} : [a_1, a_2] \leq [b_1, b_2] :\Leftrightarrow a_1 \leq b_1 \wedge a_2 \leq b_2$.
The least element of $I\!R^*$ with respect to $\leq$ is the interval $[-\infty, -\infty]$, the greatest element is $[+\infty, +\infty]$. The infimum and supremum respectively of a subset $S \subseteq I\!R^*$ are:

$$\inf_{\leq} S = [\inf_{A \in S} a_1, \inf_{A \in S} a_2], \quad \sup_{\leq} S = [\sup_{A \in S} a_1, \sup_{A \in S} a_2].$$

VIII. $\{I\!R^*, \subseteq\} : [a_1, a_2] \subseteq [b_1, b_2] :\Leftrightarrow b_1 \leq a_1 \wedge a_2 \leq b_2$.
The interval $[-\infty, +\infty]$ is the greatest element in $\{I\!R^*, \subseteq\}$, i.e. for all intervals $A \in I\!R^*$ we have $A \subseteq [-\infty, +\infty]$. But a least element is missing in $I\!R^*$. So we adjoin the empty set $[\,]$ as the least element of $I\!R^*$. The empty set $[\,]$ is a subset of any set, thus for all $A \in I\!R^*$ we have $[\,] \subseteq A$. We denote the resulting set by $\overline{I\!R^*} := I\!R^* \cup \{[\,]\}$. With this completion $\{\overline{I\!R^*}, \subseteq\}$ is a complete lattice. The infimum and supremum respectively of a subset $S \subseteq I\!R^*$ are [33, 34]:

$$\inf_{\subseteq} S = [\sup_{A \in S} a_1, \inf_{A \in S} a_2], \quad \sup_{\subseteq} S = [\inf_{A \in S} a_1, \sup_{A \in S} a_2],$$

i.e. the infimum is the intersection and the supremum is the interval (convex) hull of all intervals out of S. For $\inf_{\subseteq} S$ we shall also use the

3. Interval Arithmetic Revisited

usual symbol $\bigcap S$. $\sup_{\subseteq} S$ is occasionally written as $\bigsqcup S$. If in particular S just consists of two intervals A, B, this reads:

intersection: $A \cap B$, interval hull: $A \sqcup B$.

Since $\mathbb{R}$ is a linearly ordered set with respect to $\leq$, the interval hull is the same as the convex hull. The intersection may be empty.

In the following section we shall generalize the Interval Newton Method in such a way that for the Interval Newton Operator

$$N(X) := x - f(x)/F'(X), \quad x \in X \in I\mathbb{R} \tag{3.33}$$

the case $0 \in F'(X)$ is no longer excluded. The result of the division then can be taken from Tables 3.3 and 3.4. It is no longer an element out of $I\mathbb{R}$, but an element of the powerset $\mathbb{P}\mathbb{R}$. Thus the subtraction that occurs in (3.33) is also an operation in $\mathbb{P}\mathbb{R}$. As such it is defined by (3.29) and (3.30). As a consequence of this, of course, the operation is inclusion isotone and the inclusion property holds. We are interested in the evaluation of an expression of the form

$$Z := x - a/B, \text{ with } x, a \in \mathbb{R} \text{ and } 0 \in B \in I\mathbb{R}. \tag{3.34}$$

(3.34) can also be written as $Z = x + (-a/B)$. Multiplication of the set a/B by -1 negates and exchanges all bounds (see IV. above). Corresponding to the eight cases of Table 3.3, eight cases are again to be distinguished. The result is shown in the last column of Table 3.5.

Table 3.5. Evaluation of $Z = x - a/B$ for $x, a \in \mathbb{R}$, and $0 \in B \in I\mathbb{R}$.

	a	$B = [b_1, b_2]$	$-a/B$	$Z := x - a/B$
1	$a = 0$	$0 \in B$	$(-\infty, +\infty)$	$(-\infty, +\infty)$
2	$a \neq 0$	$B = [0, 0]$	$[\,]$	$[\,]$
3	$a < 0$	$b_1 < b_2 = 0$	$(-\infty, -a/b_1]$	$(-\infty, x - a/b_1]$
4	$a < 0$	$b_1 < 0 < b_2$	$(-\infty, -a/b_1] \cup [-a/b_2, +\infty)$	$(-\infty, x - a/b_1] \cup [x - a/b_2, +\infty)$
5	$a < 0$	$0 = b_1 < b_2$	$[-a/b_2, +\infty)$	$[x - a/b_2, +\infty)$
6	$a > 0$	$b_1 < b_2 = 0$	$[-a/b_1, +\infty)$	$[x - a/b_1, +\infty)$
7	$a > 0$	$b_1 < 0 < b_2$	$(-\infty, -a/b_2] \cup [-a/b_1, +\infty)$	$(-\infty, x - a/b_2] \cup [x - a/b_1, +\infty)$
8	$a > 0$	$0 = b_1 < b_2$	$(-\infty, -a/b_2]$	$(-\infty, x - a/b_2]$

The general rules for subtraction of the type of sets which occur in column 4 of Table 3.5, from a real number x are:

3.6 Extended Interval Arithmetic

$$x - (-\infty, +\infty) = (-\infty, +\infty),$$
$$x - (-\infty, +y] = [x - y, +\infty),$$
$$x - [y, +\infty) = (-\infty, x - y],$$
$$x - (-\infty, y] \cup [z, +\infty) = (-\infty, x - z] \cup [x - y, +\infty),$$
$$x - [\,] = [\,].$$

If in any arithmetic operation an operand is the empty set the result of the operation is also the empty set.

At the end of this Section we briefly summarize what has been developed so far.

In Section 3.3 we have considered the powerset $I\!P I\!R$ of real numbers and the subset $I\!I\!R$ of closed and bounded intervals over $I\!R$. Arithmetic operations have been defined in $I\!P I\!R$ by (3.5). We have seen that with these operations $I\!I\!R$ is an algebraically closed subset of $I\!P I\!R$ if division by an interval which contains zero is excluded.

With respect to the order relation $\leq$, $\{I\!R, \leq\}$ is a linearly ordered set. With respect to the order relation $\subseteq$, $\{I\!I\!R, \subseteq\}$ is an ordered set. Both sets $\{I\!R, \leq\}$ and $\{I\!I\!R, \subseteq\}$ are conditionally complete lattices (i.e. every non empty, bounded subset has an infimum and a supremum).

In this section we have completed the set $\{I\!R, \leq\}$ by adjoining a least element $-\infty$ and a greatest element $+\infty$. This leads to the set $I\!R^* := I\!R \cup \{-\infty\} \cup \{+\infty\}$. $\{I\!R^*, \leq\}$ then is a complete lattice (i.e. every subset has an infimum and a supremum). Similarly we have completed the set $\{I\!I\!R^*, \subseteq\}$ by adjoining the empty set $[\,]$ as a least element. This leads to the set $\overline{I\!I\!R^*} := I\!I\!R^* \cup \{[\,]\}$. $\{\overline{I\!I\!R^*}, \subseteq\}$ then is a complete lattice also.

Then we have extended interval division A/B, with $A, B \in I\!I\!R$ to the case $0 \in B$. We have seen, that division by an interval of $I\!I\!R$ which contains zero is well defined in $I\!P I\!R$ and that the result always is an element of $I\!P I\!R$, i.e. a set of real numbers.

This is an important result. We stress the fact that the result of division by an interval of $I\!I\!R$ which contains zero is not an element of $I\!I\!R$ nor of $\overline{I\!I\!R^*}$. Thus definition by an interval that contains zero does not require definition of arithmetic operations in the completed set of intervals $\overline{I\!I\!R^*} := I\!I\!R^* \cup \{[\,]\}$. Although this is often done in the literature, we have not done so here. Thus complicated definitions and rules for computing with intervals like $[-\infty, -\infty], [-\infty, 0], [0, +\infty], [+\infty, +\infty]$, and $[-\infty, +\infty]$ need not be considered. Putting aside such details makes interval arithmetic more friendly for the user.

Particular and important applications in the field of enclosure methods and validated numerics may require the definition of arithmetic operations in the complete lattice $I\!I\!R^*$. The infrequent occurrence of such applications certainly justifies leaving their realization to software and to the user. This paper aims to set out those central properties of interval arithmetic which should effectively be supported by the computer's hardware, by basic software, and by the programming languages.

∞ takes on a more subtle meaning in complex arithmetic. We defer consideration of complex arithmetic and complex interval arithmetic on the computer to a follow up paper.

3.7 The Extended Interval Newton Method

The extended Interval Newton Method can be used to compute enclosures of all zeros of a continuously differentiable function $f(x)$ in a given interval X. The iteration scheme is identical to the one defined by (3.26) in Section 3.5:

$$X_{\nu+1} := (m(X_\nu) - \frac{f(m(X_\nu))}{F'(X_\nu)}) \cap X_\nu = N(X_\nu) \cap X_\nu, \quad \nu = 0, 1, 2, \ldots,$$

with $X_0 := X$. Here again $F'(X_\nu)$ is the interval evaluation of the first derivative $f'(x)$ of the function f over the interval X_ν, and $m(X_\nu)$ is any point out of X_ν, the midpoint for example. If $f(x)$ has more than one zero in X, then the derivative $f'(x)$ has at least one zero (horizontal tangent of $f(x)$) in X also, and the interval evaluation $F'(X)$ of $f'(x)$ contains zero. Thus extended interval arithmetic has to be used to execute the Newton operator

$$N(X) = x - \frac{f(x)}{F'(X)}, \text{ with } x \in X.$$

As shown by Tables 3.3 and 3.4 the result is no longer an interval of $I\!R$. It is an element of the powerset $I\!P\!R$ which, with the exception of case 2, stretches continuously to $-\infty$ or $+\infty$ or both. The intersection $N(X) \cap X$ with the finite interval X then produces a finite set again. It may consist of a finite interval of $I\!R$, or of two separate such intervals, or of the empty set. These sets are now the starting values for the next iteration. This means that in the case where two separate intervals have occurred, the iteration has to be continued with two different starting values. This situation can occur repeatedly. On a sequential computer where only one iteration can be performed at a time all intervals which are not yet dealt with are collected in a list. This list then is treated sequentially. If more than one processor is available different subintervals can be dealt with in parallel.

Again, we illustrate this process by a simple example. The starting interval is denoted by $X = [x_1, x_2]$ and the result of the Newton operator by $N = [n_1, n_2]$. See Fig. 3.4.

Now $F'(x)$ is again a superset of all slopes of tangents of $f(x)$ in the interval $X = [x_1, x_2]$. $0 \in F'(X)$. $N(X)$ again is the set of zeros of straight lines through $(x, f(x))$ with slopes out of $F'(x)$. Let be $F'(x) = [s_1, s_2]$. Since $0 \in F'(x)$ we have $s_1 \leq 0$ and $s_2 \geq 0$. The straight lines through $(x, f(x))$

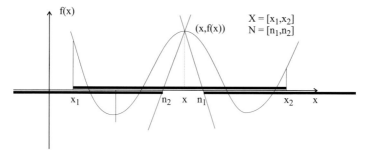

Fig. 3.4. Geometric interpretation of the extended Interval Newton Method.

with the slopes s_1 and s_2 cut the real axis in n_1 and n_2. Thus the Newton operator produces the set

$$N(X) = (-\infty, n_2] \cup [n_1, +\infty).$$

Intersection with the original set X (the former iterate) delivers the set

$$X_1 = N(X) \cap X = [x_1, n_2] \cup [n_1, x_2]$$

consisting of two finite intervals of $I\!R$. From this point the iteration has to be continued with the two starting intervals $[x_1, n_2]$ and $[n_1, x_2]$.

Remark: In case of division of a finite interval $A = [a_1, a_2]$ by an interval $B = [b_1, b_2]$ which contains zero, 8 non overlapping cases of the result were distinguished in Table 3.3 and its context. Applied to the Newton operator these 8 cases resulted in the 8 cases in Table 3.5. So far we have discussed the behaviour of the Interval Newton Method in the cases 3 to 8 of Table 3.5. We are now going to consider and interpret the particular cases 1 and 2 of Table 3.5 which, of course, may also occur. In Table 3.5 a stands for the function value and B is the enclosure of all derivatives of $f(x)$ in the interval X.

Case 2 in Table 3.5 is easy to interpret. If $B \equiv 0$ in the entire interval X then $f(x)$ is a constant in X and its value is $f(x) = a \neq 0$. So the occurrence of the empty interval in the Newton iteration indicates that the function $f(x)$ is a constant.

In case 1 of Table 3.5 the result of the Newton operator is the interval $(-\infty, +\infty)$. In this case the intersection with the former iterate X does not reduce the interval and delivers the interval X again. The Newton iteration does not converge! In this case the function value a is zero (or the numerator A in case 1 of Table 3.3 contains zero) and a zero has already been found.

In order to avoid rounding noise and to obtain safe bounds for the solution the value x may be shifted by a small ϵ to the left or right. This may transform case 1 into one of the other cases 2 to 8.

112   3. Interval Arithmetic Revisited

However, since $0 \in B$ in case 1, normally case 1 will indicate a multiple zero at the point x. This case can be further evaluated by applying the Interval Newton Method to the derivative f' of f. The values of f' as well as enclosures $F''(X)$ for the second derivative $f''(x)$ can be obtained by differentiation arithmetic (automatic differentiation) which will be dealt with in the next section.

3.8 Differentiation Arithmetic, Enclosures of Derivatives

For many applications in scientific computing it is necessary to compute the value of the derivative of a function. The Interval Newton Method requires the computation of an enclosure of the first derivative of the function over an interval. The typical "school method" first computes a formal expression for the derivative of the function by applying well known rules of differentiation. Then this expression is evaluated for a point or an interval. Differentiation arithmetic avoids the computation of a formal expression for the derivative. It computes values or enclosures of derivatives just by computing with numbers or intervals. We are now going to sketch this method for the simplest case where the value of the first derivative is to be computed. If $u(x)$ and $v(x)$ are differentiable functions then the following rules for the computation of the derivative of the sum, difference, product, and quotient of the functions are well known:

$$\begin{aligned}(u(x) \pm v(x))' &= u'(x) \pm v'(x),\\ (u(x) * v(x))' &= u'(x)v(x) + u(x)v'(x),\\ (u(x) / v(x))' &= \tfrac{1}{v^2(x)}(u'(x)v(x) - u(x)v'(x))\\ &= \tfrac{1}{v(x)}(u'(x) - \tfrac{u(x)}{v(x)}v'(x)).\end{aligned} \qquad (3.35)$$

These rules can be used to define an arithmetic for ordered pairs of numbers, similar to complex arithmetic or interval arithmetic. The first component of the pair consists of a function value $u(x_0)$ at a point x_0. The second component consists of the value of the derivative $u'(x_0)$ of the function at the point x_0. For brevity we simply write for the pair of numbers (u, u'). Then the following arithmetic for pairs follows immediately from (3.35):

$$\begin{aligned}(u, u') + (v, v') &= (u + v, u' + v'),\\ (u, u') - (v, v') &= (u - v, u' - v'),\\ (u, u') * (v, v') &= (u*v, u'v + uv'),\\ (u, u') / (v, v') &= (u/v, \tfrac{1}{v}(u' - (\tfrac{u}{v})v')), \quad v \neq 0.\end{aligned} \qquad (3.36)$$

The set of rules (3.36) is called differentiation arithmetic. It is an arithmetic which deals just with numbers. The rules (3.36) are easily programmable and are executable by a computer. These rules are now used to compute simultaneously the value and the value of the derivative of a

3.8 Differentiation Arithmetic, Enclosures of Derivatives

real function at a point x_0. For brevity we call these values the function-derivative-value-pair. Why and how can this computation be done?

Earlier in this paper we have defined a (computable) real function by an arithmetic expression in the manner that arithmetic expressions are usually defined in a programming language. Apart from the arithmetic operators $+$, $-$, $*$, and $/$, arithmetic expressions contain only three kinds of operands as basic ingredients. These are constants, variables and certain differentiable elementary functions as, for instance, $\exp, \ln, \sin, \cos, \text{sqr}, \ldots$. The derivatives of these functions are also well known.

If for a function $f(x)$ a function-derivative-value-pair is to be computed at a point x_0, all basic ingredients of the arithmetic expression of the function are replaced by their particular function-derivative-value-pair by the following rules:

$$
\begin{aligned}
\text{a constant:} \quad & c & \longrightarrow \; & (c, 0), \\
\text{the variable:} \quad & x_0 & \longrightarrow \; & (x_0, 1), \\
\text{the elementary functions:} \quad & \exp(x_0) & \longrightarrow \; & (\exp(x_0), exp(x_0)), \\
& \ln(x_0) & \longrightarrow \; & (\ln(x_0), 1/x_0), \\
& \sin(x_0) & \longrightarrow \; & (\sin(x_0), \cos(x_0)), \\
& \cos(x_0) & \longrightarrow \; & (\cos(x_0), -\sin(x_0)), \\
& \text{sqr}(x_0) & \longrightarrow \; & (\text{sqr}(x_0), 2x_0), \\
& \text{and so on.}
\end{aligned}
\qquad (3.37)
$$

Now the operations in the expression are executed following the rules (3.36) of differentiation arithmetic. The result is the function-derivative-value-pair $(f(x_0), f'(x_0))$ of the function f at the point x_0.

Example: For the function $f(x) = 25(x-1)/(x^2+1)$ the function value and the value of the first derivative are to be computed at the point $x_0 = 2$. Applying the substitutions (3.37) and the rules (3.36) we obtain

$$(f(2), f'(2)) = \frac{(25, 0)((2, 1) - (1, 0))}{(2, 1)(2, 1) + (1, 0)} = \frac{(25, 0)(1, 1)}{(4, 4) + (1, 0)} = \frac{(25, 25)}{(5, 4)} = (5, 1).$$

Thus $f(2) = 5$ and $f'(2) = 1$.

If in the arithmetic expression for the function $f(x)$ elementary functions occur in composed form, the chain rule has to be applied, for instance

$$\begin{aligned}
\exp(u(x_0)) & \longrightarrow (\exp(u(x_0)), \exp(u(x_0)) \cdot u'(x_0)) = (\exp u, u' \exp u), \\
\sin(u(x_0)) & \longrightarrow (\sin(u(x_0)), \cos(u(x_0)) \cdot u'(x_0)) = (\sin u, u' \cos u),
\end{aligned}$$

and so on.

Example: For the function $f(x) = \exp(\sin(x))$ the value and the value of the first derivative are to be computed for $x_0 = \pi$. Applying the above rules we obtain

$$(f(\pi), f'(\pi)) = (\exp(\sin(\pi)), \exp(\sin(\pi)) \cdot \cos(\pi))$$
$$= (\exp(0), -\exp(0)) = (1, -1).$$

Thus $f(\pi) = 1$ and $f'(\pi) = -1$.

Differentiation arithmetic is often called automatic differentiation. All operations are performed with numbers. A computer can easily and safely execute these operations though people cannot.

Automatic differentiation is not restricted to real functions which are defined by an arithmetic expression. Any real algorithm in essence evaluates a real expression or the value of one or several real functions. Substituting for all constants, variables and elementary functions their function-derivative-value-pair, and performing all arithmetic operations by differentiation arithmetic, computes simultaneously the function-derivative-value-pair of the result. Large program packages have been developed which do just this in particular for problems in higher dimensions.

Automatic differentiation or differentiation arithmetic simply uses the arithmetic expression or the algorithm for the function. A formal arithmetic expression or algorithm for the derivative does not explicitly occur. Of course an arithmetic expression or algorithm for the derivative is evaluated indirectly. However, this expression remains hidden. It is evaluated by the rules of differentiation arithmetic. Similarly if differentiation arithmetic is performed for a fixed interval X_0 instead of for a real point x_0, an enclosure of the range of function values and an enclosure of the range of values of the derivative over that interval X_0 are computed simultaneously. Thus, for instance, neither the Newton Method nor the Interval Newton Method requires that the user provide a formal expression for the derivatives. The derivative or an enclosure for it are computed just by use of the expression for the function itself.

Automatic differentiation allows many generalizations which all together would fill a thick book. We mention only a few of these.

If the value or an enclosure of the second derivative is needed one would use triples instead of pairs and extend the rules (3.36) for the third component by corresponding rules: $u''+v''$, $u''-v''$, $uv''+2u'v'+vu''$, ... In the arithmetic expression a constant c would now have to be replaced by the triple $(c, 0, 0)$, the variable x by $(x, 1, 0)$ and the elementary functions also by a triple with the second derivative as the third component.

Another generalization is Taylor arithmetic. It works with tuples where the first component represents the function value and the following components represent the successive Taylor coefficients. The remainder term of an integration routine for a definite integral or for an initial value problem of an ordinary differential equation usually contains a derivative of higher order. Interval Taylor arithmetic can be used to compute a safe enclosure of the remainder term over an interval. This enclosure can serve as an indicator for automatic step size control.

3.8 Differentiation Arithmetic, Enclosures of Derivatives

In arithmetics like complex arithmetic, rational arithmetic, matrix or vector arithmetic, interval arithmetic, differentiation arithmetic and Taylor arithmetic, the arithmetic itself is predefined and can be hidden in the runtime system of the compiler. The user calls the arithmetic operations by the usual operator symbols. The desired arithmetic is activated by type specification of the operands.

```
program sample;
use itaylor;
function f(x: itaylor): itaylor[lb(x)..ub(x)];
begin f := exp(5000/(sin(11+sqr(x/100))+30));
end;
var a: interval; b, fb: itaylor[0..40];
begin
        read(a);
        expand(a,b);
        fb := f(b);
        writeln ('36th Taylor coefficient: ', fb[36]);
        writeln ('40th Taylor coefficient: ', fb[40]);
end.
Test results: a = [1.001, 1.005]
36th Taylor coefficient: [-2.4139E+002, -2.4137E+002]
40th Taylor coefficient: [ 1.0759E-006, 1.0760E-006]
```

Fig. 3.5. Computation of enclosures of Taylor coefficients

As an example the PASCAL-XSC program shown in Fig. 3.5 computes and prints enclosures of the 36th and the 40th Taylor-coefficient of the function

$$f(x) = \exp(5000/(\sin(11 + \operatorname{sqr}(x/100)) + 30))$$

over the interval $a = [1.001, 1.005]$.

First the interval a is read. Then it is expanded into the 41-tuple of its Taylor coefficients $(a, 1, 0, 0, \ldots\ldots, 0)$ which is kept in b. Then the expression for $f(x)$ is evaluated in interval Taylor arithmetic and enclosures of the 36th and the 40th Taylor coefficient over the interval a are printed.

Automatic differentiation develops its full power in the case of differentiable functions of several real variables. For instance, values or enclosures of the gradient

$$\nabla f = (\frac{\partial f}{\partial x_1}, \frac{\partial f}{\partial x_2}, \ldots, \frac{\partial f}{\partial x_n})$$

of a function $f : \mathbb{R}^n \to \mathbb{R}$ or the Jacobian or Hessian matrix can be computed directly from the expression for the function f. No formal expressions for the derivatives are needed. A particular mode, the so called reverse mode, allows a

considerable acceleration for many algorithms of automatic differentiation. In the particular case of the computation of the gradient the following inequality can be shown to hold:
$$A(f, \nabla f) \leq 5A(f).$$

Here $A(f, \nabla f)$ denotes the number of operations for the computation of the gradient including the function evaluation, and $A(f)$ the number of operations for the function evaluation. For more details see [15, 16, 49].

3.9 Interval Arithmetic on the Computer

So far the basic set of all our considerations was the set of real numbers $I\!R$ or the set of extended real numbers $I\!R^* := I\!R \cup \{-\infty\} \cup \{+\infty\}$. Actual computations, however, can only be carried out on a computer. The elements of $I\!R$ and $I\!I\!R$ are in general not representable and the arithmetic operations defined for them are not executable on the computer. So we have to map these spaces and their operations onto computer representable subsets. Typical such subsets are floating-point systems, for instance, as defined by the IEEE arithmetic standard. However, in this article we do not assume any particular number representation and data format of the computer representable subsets. The considerations should apply to other data formats as well. Nevertheless, all essential properties of floating-point systems are covered.

We assume that R is a finite subset of computer representable elements of $I\!R$ with the following properties:

$$0, 1 \in R \text{ and for all } a \in R \text{ also } -a \in R.$$

The least positive non zero element of R will be denoted by L and the greatest positive element of R by G. Let be $R^* := R \cup \{-\infty\} \cup \{+\infty\}$.

Now let $\nabla : I\!R^* \to R^*$ and $\triangle : I\!R^* \to R^*$ be mappings of $I\!R^*$ onto R^* with the property that for all $a \in I\!R^*$, ∇a is the greatest lower bound of a in R^* and $\triangle a$ is the least upper bound of a in R^*. These mappings have the following three properties which also define them uniquely [33, 34]:

(R1) $\nabla a = a$ for all $a \in R^*$, $\triangle a = a$ for all $a \in R^*$,
(R2) $a \leq b \Rightarrow \nabla a \leq \nabla b$ for $a, b \in I\!R^*$, $a \leq b \Rightarrow \triangle a \leq \triangle b$ for $a, b \in I\!R^*$,
(R3) $\nabla a \leq a$ for all $a \in I\!R^*$, $a \leq \triangle a$ for all $a \in I\!R^*$.

Because of these properties ∇ is called the monotone rounding downwards and $\triangle$ is called the monotone rounding upwards. The mappings ∇ and $\triangle$ are not independent of each other. The following equalities hold for them:

$$\nabla a = -\triangle(-a) \wedge \triangle a = -\nabla(-a) \text{ for all } a \in I\!R^*.$$

3.9 Interval Arithmetic on the Computer

With the roundings $\triangledown$ and $\triangle$ arithmetic operations $\triangledown\!\!\!\!\circ$ and $\triangle\!\!\!\!\circ$, $\circ \in \{+, -, *, /\}$ can be defined in R by:

(RG) $a \triangledown\!\!\!\!\circ b := \triangledown(a \circ b)$ for all $a, b \in R$ and all $\circ \in \{+, -, *, /\}$,
$a \triangle\!\!\!\!\circ b := \triangle(a \circ b)$ for all $a, b \in R$ and all $\circ \in \{+, -, *, /\}$,
with $b \neq 0$ in case of division.

For elements $a, b \in R$ (floating-point numbers, for instance) these operations approximate the correct result $a \circ b$ in $I\!R$ by the greatest lower bound $\triangledown(a \circ b)$ and the least upper bound $\triangle(a \circ b)$ in R^* for all operations $\circ \in \{+, -, *, /\}$.

In the particular case of floating-point numbers, the IEEE arithmetic standard, for instance, requires the roundings $\triangledown$ and $\triangle$, and the corresponding operations defined by (RG). As a consequence of this all processors that provide IEEE arithmetic are equipped with the roundings $\triangledown$ and $\triangle$ and the eight operations $\triangledown\!\!\!\!+, \triangledown\!\!\!\!-, \triangledown\!\!\!\!*, \triangledown\!\!\!\!/, \triangle\!\!\!\!+, \triangle\!\!\!\!-, \triangle\!\!\!\!*,$ and $\triangle\!\!\!\!/$. On any computer each one of these roundings and operations should be provided by a single instruction which is directly supported by the computer hardware.

The IEEE arithmetic standard [78], however, separates the rounding from the arithmetic operation. First the rounding mode has to be set then one of the operations $\triangledown\!\!\!\!+, \triangledown\!\!\!\!-, \triangledown\!\!\!\!*, \triangledown\!\!\!\!/, \triangle\!\!\!\!+, \triangle\!\!\!\!-, \triangle\!\!\!\!*,$ and $\triangle\!\!\!\!/$ may be called. This slows down these operations and interval arithmetic unnecessarily and significantly.

In the preceding sections we have defined and studied the set of intervals $I\!I\!R^*$. We are now going to approximate intervals of $I\!I\!R^*$ by intervals over R^*. We consider intervals over $I\!R^*$ with endpoints in R^* of the form

$$[a_1, a_2] = \{x \in I\!R^* \mid a_1, a_2 \in R^*, a_1 \leq x \leq a_2\}.$$

The set of all such intervals is denoted by IR^*. The empty set $[\,]$ is assumed to be an element of IR^* also. Then $IR^* \subseteq I\!I\!R^*$. Note that an interval of IR^* represents a continuous set of real numbers. It is not just a set of elements of R^*! Only the bounds of intervals of IR^* are restricted to be elements of R^*.

With $I\!I\!R^*$ also the subset IR^* is an ordered set with respect to both order relations $\leq$ and $\subseteq$. It can be shown that IR^* is a complete sublattice of $I\!I\!R^*$ with respect to both order relations. For a complete proof of these properties see [33, 34]. For completeness we list the order and lattice operations in both cases. We assume that $A = [a_1, a_2]$, and $B = [b_1, b_2]$ are elements of IR^*.

$\{IR^*, \leq\}:\ [a_1, a_2] \leq [b_1, b_2] :\Leftrightarrow a_1 \leq b_1 \wedge a_2 \leq b_2.$
The least element of IR^* with respect to $\leq$ is the interval $[-\infty, -\infty]$. The greatest element is $[+\infty, +\infty]$. The infimum and supremum respectively of a subset $S \subseteq IR^*$ with respect to $\leq$ are with $A = [a_1, a_2] \in S$:

$$\inf_{\leq} S = [\inf_{A \in S} a_1, \inf_{A \in S} a_2], \quad \sup_{\leq} S = [\sup_{A \in S} a_1, \sup_{A \in S} a_2].$$

Since R^* and IR^* only contain a finite number of elements these can also be written

$$\inf_{\leq} S = [\min_{A \in S} a_1, \min_{A \in S} a_2], \quad \sup_{\leq} S = [\max_{A \in S} a_1, \max_{A \in S} a_2].$$

$\{IR^*, \subseteq\}: \quad [a_1, a_2] \subseteq [b_1, b_2] :\Leftrightarrow b_1 \leq a_1 \wedge a_2 \leq b_2$.

The least element of IR^* with respect to $\subseteq$ is the empty set $[\,]$. The greatest element is the interval $[-\infty, +\infty]$. The infimum and supremum respectively of a subset $S \in IR^*$ with respect to $\subseteq$ are with $A = [a_1, a_2] \in S$

$$\inf_{\subseteq} S = [\sup_{A \in S} a_1, \inf_{A \in S} a_2], \quad \sup_{\subseteq} S = [\inf_{A \in S} a_1, \sup_{A \in S} a_2].$$

Because of the finiteness of R^* and IR^* these can also be written

$$\inf_{\subseteq} S = [\max_{A \in S} a_1, \min_{A \in S} a_2], \quad \sup_{\subseteq} S = [\min_{A \in S} a_1, \max_{A \in S} a_2],$$

i.e. the infimum is the intersection and the supremum is the interval (convex) hull of all intervals of S. As in the case of $I\!R^*$ we shall use the usual mathematical symbols $\bigcap S$ for $inf_\subseteq S$ and $\bigcup S$ for $sup_\subseteq S$. The intersection may be empty. If in particular S consists of just two elements $A = [a_1, a_2]$ and $B = [b_1, b_2]$ this reads:

$$A \cap B = [\max(a_1, b_1), \min(a_2, b_2)] \quad \text{intersection},$$
$$A \cup B = [\min(a_1, b_1), \max(a_2, b_2)] \quad \text{interval hull}.$$

Thus, for both order relations $\leq$ and $\subseteq$ for any subset S of IR^* the infimum and supremum are the same as taken in $I\!R^*$. This is by definition the criterion for a subset of a complete lattice to be a complete sublattice. So we have the results:

$\{IR^*, \leq\}, \quad$ is a complete sublattice of $\quad \{I\!R^*, \leq\}, \quad$ and
$\{IR^*, \subseteq\}, \quad$ is a complete sublattice of $\quad \{I\!R^*, \subseteq\}$.

In many applications of interval arithmetic, it has to be determined whether an interval A is strictly included in an interval B. This is formally expressed by the notation:

$$A \subset \overset{\circ}{B}. \tag{3.38}$$

Here $\overset{\circ}{B}$ denotes the interior of B. With $A = [a_1, a_2]$ and $B = [b_1, b_2]$ (3.38) is equivalent to

$$A \subset \overset{\circ}{B} :\Leftrightarrow b_1 < a_1 \wedge a_2 < b_2.$$

In general, interval calculations are employed to determine sets that include the solution to a given problem. Since the arithmetic operations in $I\!R$ cannot in general be executed on the computer, they have to be approximated by corresponding operations in IR. These approximations are required to have the following properties:

(a) The result of any computation in IR always has to include the result of the corresponding computation in IIR.
(b) The result of the computation in IR should be as close as possible to the result of the corresponding computation in IIR.

For all arithmetic operations $\circ \in \{+,-,*,/\}$ in IIR (a) means that the computer approximation $\odot$ in IR must be defined in a way that the following inequality holds:

$$A \circ B \subseteq A \odot B \text{ for } A, B \in IR \text{ and all } \circ \in \{+,-,*,/\}. \tag{3.39}$$

Similar requirements must hold for the elementary functions. Earlier in this paper we have defined the interval evaluation of an elementary function f over an interval $A \in IIR$ by the range of function values $f(A) = \{f(a) \mid a \in A\}$. So (a) requires that for the computer evaluation $\bigcirc f(A)$ of f the following inequality holds:

$$f(A) \subseteq \bigcirc f(A) \text{ with } A \text{ and } \bigcirc f(A) \in IR. \tag{3.40}$$

(3.39) and (3.40) are necessary consequences of (a). There are reasonably good realizations of interval arithmetic on computers which only fulfil property (a).

(b) is an independent additional requirement. In the cases (3.39) and (3.40) it requires that $A \odot B$ and $\bigcirc f(A)$ should be the smallest interval in IR^* that includes the result $A \circ B$ and $f(A)$ in IIR^* respectively. It turns out that interval arithmetic on any computer is uniquely defined by this requirement. Realization of it actually is the easiest way to support interval arithmetic on the computer by hardware. To establish this is the aim of this paper.

We are now going to discuss this arithmetic in detail. First we define the mapping $\diamondsuit : IIR^* \to IR^*$ which approximates each interval A of IIR^* by its least upper bound $\diamondsuit A$ in IR^* with respect to the order relation $\subseteq$. This mapping has the property that for each interval $A = [a_1, a_2] \in IIR^*$ its image in IR^* is

(R) $\quad \diamondsuit A = \diamondsuit [a_1, a_2] = [\nabla a_1, \triangle a_2]$.

This mapping $\diamondsuit$ has the following properties which also define it uniquely [33, 34]:

(R1) $\quad \diamondsuit A = A$ for all $A \in IR^*$,
(R2) $\quad A \subseteq B \Rightarrow \diamondsuit A \subseteq \diamondsuit B$ for $A, B \in IIR^*$, $\qquad$ (monotone)
(R3) $\quad A \subseteq \diamondsuit A$ for all $A \in IIR^*$. $\qquad$ (upwardly directed)

We call this mapping $\diamondsuit$ the interval rounding. It has the additional property

(R4) $\Diamond(-A) = -\Diamond(A)$, (antisymmetry)

since with $A = [a_1, a_2]$, $-A = [-a_2, -a_1]$ and $\Diamond(-A) = [\nabla(-a_2), \Delta(-a_1)] = [-\Delta a_2, -\nabla a_1] = -[\nabla a_1, \Delta a_2] = -\Diamond A$.

The interval rounding $\Diamond : I\!I\!R^* \to I\!R^*$ is now employed in order to define arithmetic operations $\Diamondblack, \circ \in \{+, -, *, /\}$ in $I\!R$, i.e. on the computer, by

(RG) $A \Diamondblack B := \Diamond(A \circ B)$ for all $A, B \in I\!R$ and $\circ \in \{+, -, *, /\}$,
with $0 \notin B$ in case of division.

For intervals $A, B \in I\!R$ (for instance intervals the bounds of which are floating-point numbers) these operations approximate the correct result of the interval operation $A \circ B$ in $I\!I\!R$ by the least upper bound $\Diamond(A \circ B)$ in $I\!R^*$ with respect to the order relation $\subseteq$ for all operations $\circ \in \{+, -, *, /\}$.

Now we proceed similarly with the elementary functions. The interval evaluation $f(A)$ of an elementary function f over an interval $A \in I\!R$ is approximated on the computer by its image under the interval rounding $\Diamond$. Consequently the following inequality holds:

$$f(A) \subseteq \Diamond f(A) \text{ with } A \in I\!R \text{ and } \Diamond f(A) \in I\!R^*.$$

Thus $\Diamond f(A)$ is the least upper bound of $f(A)$ in $I\!R^*$ with respect to the order relation $\subseteq$.

If the arithmetic operations for elements of $I\!R$ are defined by (RG) with the rounding (R) the inclusion isotony and the inclusion property hold for the computer approximations of all interval operations $\circ \in \{+, -, *, /\}$. These are simple consequences of (R2) and (R3) respectively:

Inclusion isotony:

$$A \subseteq B \wedge C \subseteq D \Rightarrow A \circ C \subseteq B \circ D$$
$$\stackrel{(R2)}{\Rightarrow} \Diamond(A \circ C) \subseteq \Diamond(B \circ D)$$
$$\stackrel{(RG)}{\Rightarrow} A \Diamondblack C \subseteq B \Diamondblack D, \text{ for all } A, B, C, D \in I\!R.$$

Inclusion property:

$$a \in A \wedge b \in B \Rightarrow a \circ b \in A \circ B \stackrel{(R3)}{\Rightarrow} a \circ b \in \Diamond(A \circ B)$$
$$\stackrel{(RG)}{\Rightarrow} a \circ b \in A \Diamondblack B, \text{ for } a, b \in I\!R, A, B \in I\!R.$$

Both properties also hold for the interval evaluation of the elementary functions:

Inclusion isotony: $A \subseteq B \Rightarrow f(A) \subseteq f(B) \stackrel{(R2)}{\Rightarrow} \Diamond f(A) \subseteq \Diamond f(B)$,
for $A, B \in I\!R$.

Inclusion property: $a \in A \Rightarrow f(a) \in f(A) \stackrel{(R3)}{\Rightarrow} f(a) \in \Diamond f(A)$,
for $a \in \mathbb{R}, A \in I\mathbb{R}$.

Note that these two properties for the elementary functions are simple consequences of (R2) and (R3) respectively only. The optimality of the rounding $\Diamond : I\mathbb{R}^* \to I\mathbb{R}^*$ which requires that the image of an interval $A \in I\mathbb{R}^*$ is the least upper bound in $I\mathbb{R}^*$ is not necessarily required!

With these results we can define the computer evaluation of general arithmetic expressions and of real functions in interval arithmetic for an interval $X \in I\mathbb{R}$. If $f(x)$ is an arithmetic expression (consisting of constants, variables, and elementary functions connected by arithmetic operations and parentheses) an interval evaluation on the computer for an interval $X \in I\mathbb{R}$ (out of the domain of definition $D(f)$) is obtained by the following rules:

- Every constant $a \in \mathbb{R}$ is replaced by the interval $[\nabla a, \triangle a]$.
- Every occurrence of the variable x in the expression for $f(x)$ is replaced by the interval X.
- An elementary function $\varphi(x)$ is replaced by its computer evaluation $\Diamond \varphi(X)$.
- Every real operation $\circ \in \{+, -, *, /\}$ is replaced by the corresponding interval operation $\Diamond, \circ \in \{+, -, *, /\}$.
- The interval expression thus defined in interval arithmetic is evaluated on the computer.

This procedure extends the central properties of interval arithmetic — the inclusion isotony and the inclusion property — to computer evaluations of arithmetic expressions and of real functions in interval arithmetic.

(3.11) in Section 3.3 summarizes the explicit formulas (3.1), (3.2), (3.3), and (3.4) for the operations with intervals $A = [a_1, a_2]$ and $B = [b_1, b_2] \in I\mathbb{R}$ by

$$A \circ B = [\min_{i,j=1,2}(a_i \circ b_j), \max_{i,j=1,2}(a_i \circ b_j)] \text{ for all } \circ \in \{+, -, *, /\} \text{ with } 0 \notin B \text{ in case of division.}$$

Thus, the definition of the operations in $I\mathbb{R}$ by (RG) and of the interval rounding $\Diamond : I\mathbb{R}^* \to I\mathbb{R}^*$ by (R) leads directly to the following formula for the operations for intervals $A = [a_1, a_2]$ and $B = [b_1, b_2] \in I\mathbb{R}$:

$$A \Diamond B := \Diamond(A \circ B) = [\nabla \min_{i,j=1,2}(a_i \circ b_j), \triangle \max_{i,j=1,2}(a_i \circ b_j)], \\ \circ \in \{+, -, *, /\}, \text{ with } 0 \notin B \text{ in case of division.} \quad (3.41)$$

Since $\nabla : I\mathbb{R}^* \to \mathbb{R}^*$ and $\triangle : I\mathbb{R}^* \to \mathbb{R}^*$ are monotone mappings (R2), we obtain

$$A \Diamond B := \Diamond(A \circ B) = [\min_{i,j=1,2}(a_i \nabla b_j), \max_{i,j=1,2}(a_i \triangle b_j)], \\ \circ \in \{+, -, *, /\}, \text{ with } 0 \notin B \text{ in case of division.} \quad (3.42)$$

Employing this equation and the explicit formulas for the arithmetic operations in $I\!R$ listed under I, II, III, IV, V, VI, in Section 3.6 leads to the following formulas for the execution of the arithmetic operations $\diamondsuit, \circ \in \{+, -, *, /\}$, in IR on the computer for intervals $A = [a_1, a_2]$ and $B = [b_1, b_2] \in IR$:

I. Equality: $[a_1, a_2] = [b_1, b_2] :\Leftrightarrow a_1 = b_1, a_2 = b_2$.
II. Addition: $[a_1, a_2] \diamondsuit [b_1, b_2] := [a_1 \triangledown b_1, a_2 \triangle b_2]$.
III. Subtraction: $[a_1, a_2] \diamondsuit [b_1, b_2] := [a_1 \triangledown b_2, a_2 \triangle b_1]$.
IV. Negation: $A = [a_1, a_2], -A = [-a_2, -a_1]$.
V. Multiplication: see Table 3.6.
VI. Division, $0 \notin B$: see Table 3.7.

Table 3.6. Multiplication of two intervals $A, B \in IR$ on the computer.

$A \diamondsuit B$	$b_1 \geq 0$	$b_1 < 0 < b_2$	$b_2 \leq 0$
$a_1 \geq 0$	$[a_1 \triangledown b_1, a_2 \triangle b_2]$	$[a_2 \triangledown b_1, a_2 \triangle b_2]$	$[a_2 \triangledown b_1, a_1 \triangle b_2]$
$a_1 < 0 < a_2$	$[a_1 \triangledown b_2, a_2 \triangle b_2]$	$[\min(a_1 \triangledown b_2, a_2 \triangledown b_1),$ $\max(a_1 \triangle b_1, a_2 \triangle b_2)]$	$[a_2 \triangledown b_1, a_1 \triangle b_1]$
$a_2 \leq 0$	$[a_1 \triangledown b_2, a_2 \triangle b_1]$	$[a_1 \triangledown b_2, a_1 \triangle b_1]$	$[a_2 \triangledown b_2, a_1 \triangle b_1]$

Table 3.7. Division of two intervals $A, B \in IR$ with $0 \notin B$ on the computer.

$A \diamondsuit B$	$b_1 > 0$	$b_2 < 0$
$a_1 \geq 0$	$[a_1 \triangledown b_2, a_2 \triangle b_1]$	$[a_2 \triangledown b_2, a_1 \triangle b_1]$
$a_1 < 0 < a_2$	$[a_1 \triangledown b_1, a_2 \triangle b_1]$	$[a_2 \triangledown b_2, a_1 \triangle b_2]$
$a_2 \leq 0$	$[a_1 \triangledown b_1, a_2 \triangle b_2]$	$[a_2 \triangledown b_1, a_1 \triangle b_2]$

These formulas show, in particular, that the operations $\diamondsuit, \circ \in \{+, -, *, /\}$, in IR are executable on a computer if the operations $\triangledown$ and $\triangle, \circ \in \{+, -, *, /\}$, for elements of R are available. These operations have been defined earlier in this Section by

(RG) $a \triangledown b := \triangledown(a \circ b)$ and $a \triangle b := \triangle(a \circ b)$ for $a, b \in R$ and $\circ \in \{+, -, *, /\}$
with $b \neq 0$ in case of division.

This in turn shows the importance of the roundings $\triangledown : I\!R^* \to R^*$ and $\triangle : I\!R^* \to R^*$.

Table 3.8. The 8 cases of the division of two intervals $A \diamond B$, with $A, B \in I\!R$ and $0 \in B$.

case	$A = [a_1, a_2]$	$B = [b_1, b_2]$	$A \diamond B$
1	$0 \in A$	$0 \in B$	$[-\infty, +\infty]$
2	$0 \notin A$	$B = [0, 0]$	$[\,]$
3	$a_2 < 0$	$b_1 < b_2 = 0$	$[a_2 \nabla b_1, +\infty]$
4	$a_2 < 0$	$b_1 < 0 < b_2$	$[-\infty, a_2 \triangle b_2] \cup [a_2 \nabla b_1, +\infty]$
5	$a_2 < 0$	$0 = b_1 < b_2$	$[-\infty, a_2 \triangle b_2]$
6	$a_1 > 0$	$b_1 < b_2 = 0$	$[-\infty, a_1 \triangle b_1]$
7	$a_1 > 0$	$b_1 < 0 < b_2$	$[-\infty, a_1 \triangle b_1] \cup [a_1 \nabla b_2, +\infty]$
8	$a_1 > 0$	$0 = b_1 < b_2$	$[a_1 \nabla b_2, +\infty]$

In case of division by an interval B which contains zero, eight cases had to be distinguished in Table 3.3. On the computer these cases have to be performed as shown in Table 3.8. With $A = [a_1, a_2]$ and $B = [b_1, b_2]$ Table 3.9 shows the same cases as Table 3.8 in another representation.

Table 3.9. The result of the division $A \diamond B$, with $A, B \in I\!R$ and $0 \in B$.

$A \diamond B$	$B = [0, 0]$	$b_1 < b_2 = 0$	$b_1 < 0 < b_2$	$0 = b_1 < b_2$
$a_2 < 0$	$[\,]$	$[a_2 \nabla b_1, +\infty]$	$[-\infty, a_2 \triangle b_2] \cup [a_2 \nabla b_1, +\infty]$	$[-\infty, a_2 \triangle b_2]$
$a_1 \leq 0 \leq a_2$	$[-\infty, +\infty]$	$[-\infty, +\infty]$	$[-\infty, +\infty]$	$[-\infty, +\infty]$
$a_1 > 0$	$[\,]$	$[-\infty, a_1 \triangle b_1]$	$[-\infty, a_1 \triangle b_1] \cup [a_1 \nabla b_2, +\infty]$	$[a_1 \nabla b_2, +\infty]$

The generalized Newton operator requires the subtraction of a set which tends to plus or minus infinity or both or which is the empty set from a real number x. On the computer the corresponding rules now appear in the form

$$x \diamond [-\infty, +\infty] = [-\infty, +\infty],$$
$$x \diamond [-\infty, y] = [x \nabla y, +\infty],$$
$$x \diamond [y, +\infty] = [-\infty, x \triangle y],$$
$$x \diamond ([-\infty, y] \cup [z, +\infty]) = [-\infty, x \triangle z] \cup [x \nabla y, +\infty],$$
$$x \diamond [\,] = [\,].$$

After the computation of the Interval Newton Operator the intersection with a finite interval $[c_1, c_2]$ still has to be taken in the generalized Interval Newton Method. The result may be one or two finite intervals or the empty interval $[\,]$. These cases are expressed by the following explicit formulas:

$$(x \Diamond [-\infty, +\infty]) \cap [c_1, c_2] = [c_1, c_2],$$
$$(x \Diamond [-\infty, y]) \cap [c_1, c_2] = [x \nabla y, c_2] \text{ or } [\,],$$
$$(x \Diamond [y, +\infty]) \cap [c_1, c_2] = [c_1, x \triangle y] \text{ or } [\,],$$
$$x \Diamond ([-\infty, y] \cup [z, +\infty]) \cap [c_1, c_2] = [c_1, x \triangle z] \cup [x \nabla y, c_2] \text{ or } [\,],$$
$$(x \Diamond [\,]) \cap [c_1, c_2] = [\,] \cap [c_1, c_2] = [\,].$$

For geometric reasons $[c_1, c_2]$ can only occur as the result of the intersection in the first case.

For interval arithmetic the roundings $\nabla : I\!R^* \to R^*$ and $\triangle : I\!R^* \to R^*$ are of particular interest. They can be defined by the following properties:

$$\nabla a := \max\{x \in R^* \mid x \leq a\}, \quad \text{monotone rounding downwards, and}$$
$$\triangle a := \min\{x \in R^* \mid x \geq a\}, \quad \text{monotone rounding upwards.}$$

The following equalities hold for ∇ and $\triangle$:

$$\nabla a = -\triangle(-a), \quad \text{and} \quad \triangle a = -\nabla(-a),$$

i.e. they can be expressed by one another.

For completeness we give an explicit description of the rounding $\nabla : I\!R^* \to R^*$ in the case that R^* is a floating-point system. A floating-point number is a real number of the form $x = m \cdot b^e$. Here m is the mantissa, b is the base of the number system in use and e is the exponent. b is an integer greater than one. The exponent is an integer between two fixed integer bounds $e1, e2$, and in general $e1 \leq 0 \leq e2$. The mantissa is of the form $m = \circ \sum_{i=1}^{r} d[i] \cdot b^{-i}$. Here $\circ \in \{+, -\}$ is the sign of the number. The $d[i]$ are the digits of the mantissa. In a normalized floating-point system they have the property $d[i] \in \{0, 1, \ldots b - 1\}$, for all $i = 1(1)r$ and $d[1] \neq 0$. Thus $|m| < 1$. Without the condition $d[1] \neq 0$, floating-point numbers are said to be unnormalized. The set of normalized floating-point numbers does not contain zero. So zero is adjoined to R^*. For a unique representation of zero it is often assumed that $m = 0, 00\ldots 0$ and $e = 0$. A floating-point system thus depends on the constants $b, r, e1$, and $e2$. Here we denote it by $R = R(b, r, e1, e2)$, then $R^* := R \cup \{-\infty\} \cup \{+\infty\}$.

In the following description of $\nabla : I\!R^* \to R^*$ we use the abbreviation $G := 0.(b-1)(b-1)\ldots(b-1) \cdot b^{e2}$ for the greatest positive floating-point number. Then we obtain for ∇a:

$$\nabla a = \begin{cases} +\infty & \text{for } a = +\infty, \\ +G & \text{for } +G \leq a < +\infty, \\ +0.a[1]a[2]\ldots a[r] \cdot b^e & \text{for } b^{e1-1} \leq a < +G, \\ +0.000\ldots0 \cdot b^0 & \text{for } 0 \leq a < b^{e1-1}, \\ -0.100\ldots0 \cdot b^{e1} & \text{for } -b^{e1-1} \leq a < 0, \\ -0.a[1]a[2]\ldots a[r] \cdot b^e & \text{for } -G \leq a < -b^{e1-1} \wedge \\ & \quad a[r+i] = 0 \text{ for all } i \geq 1, \\ -0.100\ldots0 \cdot b^{e+1} & \text{for } -G \leq a < -b^{e1-1} \wedge \\ & \quad a[i] = b-1 \text{ for all } i = 1(1)r \wedge \\ & \quad a[r+i] \neq 0 \text{ for any } i \geq 1, \\ -(0.a[1]a[2]\ldots a[r] + b^{-r}) \cdot b^e & \text{for } -G \leq a < -b^{e1-1} \wedge \\ & \quad a[i] \neq b-1 \text{ for any } i \in \{1, 2, \ldots, r\} \wedge \\ & \quad a[r+i] \neq 0 \text{ for any } i \geq 1, \\ -\infty & \text{for } -\infty \leq a < -G. \end{cases}$$

Using the function $[a]$ (the greatest integer less than or equal to a) the description of ∇a can be shortened:

$$\nabla a = \begin{cases} +\infty & \text{for } a = +\infty, \\ +G & \text{for } +G \leq a < +\infty, \\ [m_\infty \cdot b^r] \cdot b^{-r} \cdot b^e & \text{for } b^{e1-1} \leq |a| \leq +G, \\ 0.00\ldots0 \cdot b^0 & \text{for } 0 \leq a < b^{e1-1}, \\ -0.100\ldots0 \cdot b^{e1} & \text{for } -b^{e1-1} \leq a < 0, \\ -\infty & \text{for } -\infty \leq a < -G. \end{cases}$$

The more detailed description of ∇a above shows that a normalization may still be necessary.

A few additional but very similar cases occur if in the case $e < e1$, e is set to $e1$ and unnormalized mantissas are permitted.

In these representations for ∇a we have assumed that a floating-point number is represented by the so called sign-magnitude representation. For real numbers $a \geq 0$ the rounded value ∇a is obtained by truncation of a after the rth digit of the normalized mantissa m_∞ of a. If we denote this process by $t(a)$ (truncation), we have

$$\nabla a = t(a) \quad \text{for} \quad a \geq 0.$$

This is very easy to execute. Truncation can also be used to perform the rounding ∇a in case of negative numbers $a < 0$ if negative numbers are represented by their b-complement. Then the rounded value ∇a can be obtained by truncation of the b-complement $a + x$ of a via the process:

$$\nabla a = t(a + x) - x \quad \text{for} \quad a \geq 0, \tag{3.43}$$

with a suitable x. See Fig. 3.6.

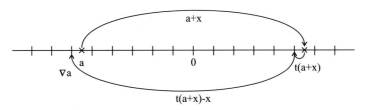

Fig. 3.6. Execution of the rounding ∇a in case of b-complement representation of negative numbers $a < 0$.

Example: We assume that the decimal number system is used, and that the mantissa has three decimal digits. Then we obtain for the positive real number $a = 0.354672 \cdot 10^3 \in \mathbb{R}$:

$$\nabla a = t(a) = 0.354 \cdot 10^3.$$

For the negative real number $a = -0.354672 \cdot 10^3$ we obtain obviously

$$\nabla a = -0.355 \cdot 10^3.$$

This value is obtained by application of (3.43) with $x = 1.00....0 \cdot 10^3$:

$$a + x = 0.645328 \cdot 10^3,$$
$$t(a + x) = 0.645 \cdot 10^3,$$
$$\nabla a = t(a + x) - x = 0.355 \cdot 10^3.$$

Here the easily executable b-complement has been taken twice. In between the function $t(a)$ was applied which also is easily executable. These three steps are particularly simple if the binary number system is used.

It is interesting that in case of the $(b - 1)$-complement representation of negative numbers the monotone rounding downwards ∇a cannot be executed by the function $t(a)$. This representation is isomorphic to the sign-magnitude representation.

In the preceding Sections 3.1 to 3.8 ideal interval arithmetic for elements of $I\mathbb{R}$ including division by an interval which contains zero has been developed. In no case did the symbols $-\infty$ and $+\infty$ occur as result of an interval operation. This is not so in this Section where interval arithmetic on the computer is considered. Here $-\infty$ and $+\infty$ can occur as result of the roundings ∇ and $\triangle$, and as result of the operations $a \nabla b$ and $a \triangle b$, $\circ \in \{+, -, *, /\}$, respectively. The interval rounding is defined by $\Diamond A := [\nabla a_1, \triangle a_2]$, and the arithmetic operations for intervals $A, B \in I\mathbb{R}$ are defined by $A \Diamond B := \Diamond (A \circ B)$, $\circ \in \{+, -, *, /\}$. As a consequence of this the symbols $-\infty$ and $+\infty$ can also occur as bounds of the result of an interval operation.

This happens, for instance, in case of division by an interval which contains zero, see Table 3.8. The extended Interval Newton Method is an example of this. We have studied this process in detail. Here very large intervals with $-\infty$ and $+\infty$ as bounds only appear intermediately. They disappear again as soon as the intersection with the previous approximation is taken. Finally the diameters of the approximations decrease to small bounds for the solution.

Among the six interval operations addition, subtraction, multiplication, division, intersection, and interval hull, the intersection is the only operation which can reduce an interval which stretches to $-\infty$ or $+\infty$ or both to a finite interval again. This step is advantageously used in the extended Interval Newton Method.

Also certain elementary functions can reduce an interval which stretches to $-\infty$ or $+\infty$ or both to a finite interval again. In such a case continuation of the computation may also be reasonable. The user has to take care that such situations are appropriately treated in his program.

In general, the appearance of $-\infty$ or $+\infty$ in the result of an interval operation indicates that an exponent overflow has occurred or that an operation or an elementary function has been called outside its range of definition. This means that the computation has gotten out of control. In this case continuation of the computation is not really recommendable. An appropriate scaling of the problem may be necessary.

Here the situation is very different from a conventional floating-point computation. In floating-point arithmetic the general directive often is just to "compute" at any price, hoping that at the end of the computation something that is reasonable will be delivered. In this process the non numbers $-\infty$, $+\infty$, and even NaN (not a number) are often treated as numbers and the computation is continued with these entities. Since a floating-point computation often flips out of control anyhow it must be the user's responsibility to control and judge the final result by other means.

In interval mathematics the general philosophy is very different. The user and the computation itself are controlling the computational process at any time. In general, an interval computation is aiming to compute small bounds for the solution of the problem. If during a computation the intervals grow overly large or even an interval appears which stretches to $-\infty$ or $+\infty$ or both, this should be taken as a severe warning. It should cause the user to think about and study the computational process again with the aim of obtaining smaller intervals. Blind continuation of the computation even with non numbers as in the case of floating-point arithmetic hoping that something reasonable will come out at the end is in strong contradiction to the philosophy and basic understanding of interval mathematics.

3.10 Hardware Support for Interval Arithmetic

An interval operation requires the computation of the lower and the upper bound of the resulting interval. For the four basic operations each of these bounds can be expressed by a single floating-point operation with particular bounds of the interval operands. The lower bound of the resulting interval has to be computed with rounding downwards and the upper bound with rounding upwards. While addition and subtraction are straightforward, multiplication and division require a detailed case analysis and in case of multiplication additionally a maximum / minimum computation if both interval operands contain zero. This may slow down these operations considerably in particular if the case analysis is performed in software. Thus in summary an interval operation is slower by a factor of at least two on a conventional sequential processor in comparison with the corresponding floating-point operation.

We show in this section that with dedicated hardware interval arithmetic can be made more or less as fast as simple floating-point arithmetic. The cost increase for the additional hardware is relatively modest and it is close to zero on superscalar processors. Although different in detail we follow in this Section ideas of [71, 72].

We assume in this section that one arithmetic operation as well as one comparison cost one unit of time whereas switches controlled by one bit as the sign bit or data transports inside the unit are free of charge. For simplicity we denote the computer operations for intervals in this section by $+, -, *,$ and $/$. The interval operands are denoted by $A = [a_1, a_2]$ and $B = [b_1, b_2]$. The lower bound, upper bound respectively of the result is denoted by lb, ub respectively, i.e. $[lb, ub] := [a_1, a_2] \circ [b_1, b_2], \circ \in \{+, -, *, /\}$.

3.10.1 Addition $A + B$ and Subtraction $A - B$

The formulas for addition and subtraction

$$[a_1, a_2] + [b_1, b_2] = [a_1 \triangledown b_1, a_2 \triangle b_2],$$
$$[a_1, a_2] - [b_1, b_2] = [a_1 \triangledown b_2, a_2 \triangle b_1]$$

require no conditionals or exceptions. They show that in comparison with floating-point arithmetic a time factor of 2 is achieved with one arithmetic unit. Duplication of this unit yields a factor of 1. In the case of addition we have with one arithmetic unit sequentially:

(A) $lb := a_1 \triangledown b_1$;
(B) $ub := a_2 \triangle b_2$;

and with two arithmetic units in parallel:

(C) $lb := a_1 \triangledown b_1$; $ub := a_2 \triangle b_2$;

3.10.2 Multiplication $A * B$

A basic method for the multiplication of two intervals is the method of case distinction. Nine cases have been distinguished in Table 3.6. In eight of the nine cases one multiplication with directed rounding suffices for the computation of each bound of the resulting interval. When both interval operands contain zero as an interior point two multiplications with directed roundings and one comparison have to be performed for each bound of the resulting interval. The case selection depends on the sign bits of the interval operands. It may be performed by hardware multiplexers which select one of two inputs. On a sequential processor with one multiplier and one comparator the following algorithm solves the problem:

Algorithm 1:

The eight cases with only one multiplication for each bound can be obtained by:

(A) $lb := $ if $(b_1 \geq 0 \vee (a_2 \leq 0 \wedge b_2 > 0))$ then a_1 else a_2
    $\triangledown$ if $(a_1 \geq 0 \vee (a_2 > 0 \wedge b_2 \leq 0))$ then b_1 else b_2;
(B) $ub := $ if $(b_1 \geq 0 \vee (a_1 \geq 0 \wedge b_2 > 0))$ then a_2 else a_1
    $\triangle$ if $(a_1 \geq 0 \vee (a_2 > 0 \wedge b_1 \geq 0))$ then b_2 else b_1;

and the final case where two multiplications have to be performed for each bound by:

(C) $p := a_1 \triangledown b_2$;
(D) $q := a_2 \triangledown b_1$;
(E) $lb := \min(p, q);\ r := a_1 \triangle b_1$;
(F) $\qquad\qquad s := a_2 \triangle b_2$;
(G) $\qquad\qquad ub := \max(r, s)$;

Taking all parts together we have:
if $(a_1 < 0 \wedge a_2 > 0 \wedge b_1 < 0 \wedge b_2 > 0)$ **then**
    $\{(C),(D),(E),(F),(G)\}$
else
    $\{(A),(B)\}$;

The correctness of the algorithm can be checked against the case distinctions of Table 3.6. The algorithm needs 5 time steps in the worst case. In all the other cases the product can be computed in two time steps.

If two multipliers and one comparator are provided the same algorithm reduces the execution time to one time step for (A), (B) and three time steps for (C), (D), (E), (F), (G). Two multiplications and a comparison can then be performed in parallel:

Algorithm 2:

(A) $lb := $ if $(b_1 \geq 0 \vee (a_2 \leq 0 \wedge b_2 > 0))$ then a_1 else a_2
    $\triangledown$ if $(a_1 \geq 0 \vee (a_2 > 0 \wedge b_2 \leq 0))$ then b_1 else b_2;
    $ub := $ if $(b_1 \geq 0 \vee (a_1 \geq 0 \wedge b_2 > 0))$ then a_2 else a_1
    $\triangle$ if $(a_1 \geq 0 \vee (a_2 > 0 \wedge b_1 \geq 0))$ then b_2 else b_1;
    and

(B) $p := a_1 \triangledown b_2; \quad q := a_2 \triangledown b_1;$
(C) $lb := \min(p, q);$ $\quad\quad\quad\quad\quad\quad\quad r := a_1 \triangle b_1; \quad s := a_2 \triangle b_2;$
(D) $\quad\quad\quad\quad\quad\quad\quad\quad\quad\quad\quad\quad\quad\quad ub := \max(r, s);$

if $(a_1 < 0 \wedge a_2 > 0 \wedge b_1 < 0 \wedge b_2 > 0)$ **then** $\{(B),(C),(D)\}$ **else** (A);

The resulting interval is delivered either in step (A) or in step (C) (minimum) and step (D) (maximum). In step (A) one multiplication for each bound suffices while in the steps (B), (C), (D) a second multiplication and a comparison are necessasry for each bound. This case where both operands contain zero occurs rather rarely. So the algorithm shows that on a processor with two multipliers and one comparator an interval multiplication can in general be performed in the same time as a floating-point multiplication.

There are applications where a large number of interval products have to be computed consecutively. This is the case, for instance, if the scalar product of two interval vectors or a matrix vector product with interval components is to be computed. In such a case it is desirable to perform the computation in a pipeline. Algorithms like 1 and 2 can, of course, be performed in a pipeline. But in these algorithms the time needed to compute an interval product heavily depends on the data. In algorithm 1 the computation of an interval product requires 2 or 5 time steps and in algorithm 2, 1 or 3 time steps. So the pipeline would have to provide 5 time steps in case of algorithm 1 and 3 in case of algorithm 2 for each interval product. I.e. the worst case rules the pipeline. The pipeline can not easily draw advantage out of the fact that in the majority of cases the data would allow to compute the product in 2 or 1 time step, respectively.

There are other methods for computing an interval product which, although they look more complicated at first glance, lead to a more regular pipeline. These methods compute an interval product in the same number of time steps as algorithms 1 and 2. The following two algorithms display such possibilities.

Algorithm 3:

By (9.4) the interval product can be computed by the following formula:

$$A * B := [\triangledown \min(a_1 * b_1, a_1 * b_2, a_2 * b_1, a_2 * b_2),$$
$$\triangle \max(a_1 * b_1, a_1 * b_2, a_2 * b_1, a_2 * b_2)].$$

This leads to the following 5 time steps for the computation of $A * B$ using 1 multiplier, 2 comparators and 2 assignments:

(A) $p := a_1 * b_1;$
(B) $q := a_1 * b_2;$
(C) $r := a_2 * b_1; \text{ MIN} := \min(p, q); \text{ MAX} := \max(p, q);$
(D) $s := a_2 * b_2; \text{ MIN} := \min(\text{MIN}, r); \text{ MAX} := \max(\text{MAX}, r);$
(E) $\quad\quad\quad lb := \triangledown \min(\text{MIN}, s); \quad ub := \triangle \max(\text{MAX}, s);$

Note that here the minimum and maximum are taken from the unrounded products of double length. The algorithm always needs 5 time steps. In algorithm 1 this is the worst case.

Algorithm 4:
Using the same formula but 2 multipliers, 2 comparators and 2 assignments leads to:

(A) $p := a_1 * b_1; q := a_1 * b_2;$
(B) $r := a_2 * b_1; s := a_2 * b_2;$ $\text{MIN} := \min(p, q); \text{MAX} := \max(p, q);$
(C) $\text{MIN} := \min(\text{MIN}, r); \text{MAX} := \max(\text{MAX}, r);$
(D) $lb := \nabla \min(\text{MIN}, s); ub := \triangle \max(\text{MAX}, s);$

Again the minimum and maximum are taken from the unrounded products. The algorithm needs 4 time steps. This is one time step more than the corresponding algorithm 2 using case distinction with two multipliers.

3.10.3 Interval Scalar Product Computation

Let us denote the components of the two interval vectors $A = (A_k)$ and $B = (B_k)$ by $A_k = [a_{k1}, a_{k2}]$ and $B_k = [b_{k1}, b_{k2}]$, $k = 1(1)n$. Then the product $A_k * B_k$ is to be computed by

$$A_k * B_k = [\min_{i,j=1,2}(a_{ki} * b_{kj}), \max_{i,j=1,2}(a_{ki} * b_{kj})], \quad k = 1(1)n.$$

The formula for the scalar product now reads:

$$[lb, ub] := A \diamondsuit B := \diamondsuit (A * B) := \diamondsuit \sum_{k=1}^{n} A_k * B_k$$

$$:= [\nabla \sum_{k=1}^{n} \min_{i,j=1,2}(a_{ki} * b_{kj}), \triangle \sum_{k=1}^{n} \max_{i,j=1,2}(a_{ki} * b_{kj})].$$

This leads to the following pipeline using 1 multiplier, 2 comparators, and 2 long fixed-point accumulators (see chapter 1):

Algorithm 5:
(A) $p := a_{k1} * b_{k1};$
(B) $q := a_{k1} * b_{k2};$
(C) $r := a_{k2} * b_{k1};$ $\quad \text{MIN} := \min(p, q); \text{MAX} := \max(p, q);$
(D) $s := a_{k2} * b_{k2};$ $\quad \text{MIN} := \min(\text{MIN}, r); \text{MAX} := \max(\text{MAX}, r);$
(E) $p := a_{k+1,1} * b_{k+1,1};$ $\quad \text{MIN} := \min(\text{MIN}, s); \text{MAX} := \max(\text{MAX}, s);$
(F) $q := a_{k+1,1} * b_{k+1,2};$ $\quad lb := lb + \text{MIN}; ub := ub + \text{MAX};$

(G) $r := a_{k+1,2} * b_{k+1,1};$ MIN $:= \min(p, q);$ MAX $:= \max(p, q);$
(H) $s := a_{k+1,2} * b_{k+1,2};$ MIN $:= \min(\text{MIN}, r);$ MAX $:= \max(\text{MAX}, r);$
$\phantom{(H) s := a_{k+1,2} * b_{k+1,2};}$ MIN $:= \min(\text{MIN}, s);$ MAX $:= \max(\text{MAX}, s);$
$\phantom{(H) s := a_{k+1,2} * b_{k+1,2};}$ $lb := lb + \text{MIN};$ $ub := ub + \text{MAX};$

....

$lb := \nabla(lb + \text{MIN});$ $ub := \triangle(ub + \text{MAX});$

This algorithm shows that in each sequence of 4 time steps one interval product can be accumulated. Again the minimum and maximum are taken from the unrounded products. Only at the very end of the accumulation of the bounds is a rounding applied. Then lb and ub are floating-point numbers which optimally enclose the product $A*B$ of the two interval vectors A and B.

In the algorithms 3, 4, and 5 the unrounded, double length products were compared and used for the computation of their minimum and maximum corresponding to (3.41). This requires comparators of double length. This can be avoided if formula (3.42) is used instead:

$$A * B := [\min(a_1 \nabla b_1, a_1 \nabla b_2, a_2 \nabla b_1, a_2 \nabla b_2),$$
$$\max(a_1 \triangle b_1, a_1 \triangle b_2, a_2 \triangle b_1, a_2 \triangle b_2)].$$

Computation of the 8 products $a_i \nabla b_j, a_i \triangle b_j, i, j = 1, 2$, can be avoided if the exact flag of the IEEE arithmetic is used. In general $a \nabla b$ and $a \triangle b$ differ only by one unit in the last place and we have

$$a \nabla b \leq a * b \leq a \triangle b.$$

If the computation of the product $a * b$ leads already to a floating-point number which needs no rounding, then the product is called exact and we have:

$$a \nabla b = a * b = a \triangle b.$$

If the product $a * b$ is not a floating-point number, then it is "not exact" and the product with rounding upwards can be obtained by taking the successor $a \triangle b := \text{succ}(a \nabla b)$. This changes algorithm 4, for instance, into

Algorithm 6:

(A) $p := a_1 \nabla b_1; q := a_1 \nabla b_2;$
(B) $r := a_2 \nabla b_1; s := a_2 \nabla b_2;$ MIN $:= \min(p, q);$ MAX $:= \max(p, q);$
(C) MIN $:= \min(\text{MIN}, r);$ MAX $:= \max(\text{MAX}, r);$
(D) $lb := \min(\text{MIN}, s);$ MAX $:= \max(\text{MAX}, s);$
(E) if MAX = "exact" then $ub := \text{MAX}$ else $ub := succ(\text{MAX});$

The algorithm requires one additional step in comparison with algorithm 4 where products of double length have been compared.

3.10.4 Division A / B

If $0 \notin B$, 6 different cases have been distinguished as listed in Table 3.7. If $0 \in B$, 8 cases have to be considered. These are listed in Table 3.8. This leads directly to the following.

Algorithm 7:
if $b_2 < 0 \vee b_1 > 0$ then
{
　　$lb := ($ if $b_1 > 0$ then a_1 else $a_2) \triangledown$
　　　　$($ if $a_1 \geq 0 \vee (a_2 > 0 \wedge b_2 < 0)$ then b_2 else $b_1)$;
　　$ub := ($ if $b_1 > 0$ then a_2 else $a_1) \triangle$
　　　　$($ if $a_1 \geq 0 \vee (a_2 > 0 \wedge b_1 > 0)$ then b_1 else $b_2)$;

} else {
if $(a_1 \leq 0 \wedge 0 \leq a_2 \wedge b_1 \leq 0 \wedge 0 \leq b_2)$ then $[lb, ub] := [-\infty, +\infty]$;
if $((a_2 < 0 \vee a_1 > 0) \wedge b_1 = 0 \wedge b_2 = 0)$ then $[lb, ub] := [\,]$;
if $(a_2 < 0 \wedge b_2 = 0)$ then $[lb, ub] := [a_2 \triangledown b_1, +\infty]$;
if $(a_2 < 0 \wedge b_1 = 0)$ then $[lb, ub] := [-\infty, a_2 \triangle b_2]$;
if $(a_1 > 0 \wedge b_2 = 0)$ then $[lb, ub] := [-\infty, a_1 \triangle b_1]$;
if $(a_1 > 0 \wedge b_1 = 0)$ then $[lb, ub] := [a_1 \triangledown b_2, +\infty]$;
if $(a_2 < 0 \wedge b_1 < 0 \wedge b_2 > 0)$ then { $[lb_1, ub_1] := [-\infty, a_2 \triangle b_2]$;
　　　　　　　　　　　　　　　　　　　　　$[lb_2, ub_2] := [a_2 \triangledown b_1, +\infty]$; }
if $(a_1 > 0 \wedge b_1 < 0 \wedge b_2 > 0)$ then { $[lb_1, ub_1] := [-\infty, a_1 \triangle b_1]$;
　　　　　　　　　　　　　　　　　　　　　$[lb_2, ub_2] := [a_1 \triangledown b_2, +\infty]$; }
}

The algorithm is organized in such a way that the most complicated cases, where the result consists of two separate intervals, appear at the end. It would be possible also in these cases to write the result as a single interval which then would overlap the point infinity. In such an interval the lower bound would then be greater than the upper bound. This could cause difficulties with the order relation. So we prefer the notation with the two separate intervals. On the other hand, the representation of the result as an interval which overlaps the point infinity has advantages as well. The result of an interval division then always consists of just two bounds. In the Newton step the separation into two intervals then would have to be done by the intersection.

In practice, division by an interval that contains zero occurs infrequently. So algorithm 7 shows again that on a processor with two dividers and some multiplexer equipment an interval division can in general be performed in the same time as a floating-point division.

Variants of the algorithms discussed in this Section can, of course, also be used. In algorithm 7, for instance, the sequence of the if-statements after the *else* could be interchanged. If between these if-statements all semicolons are replaced by an *else* the result may be obtained faster.

3.10.5 Instruction Set for Interval Arithmetic

Convenient high level programming languages with particular data types and operators for intervals, the XSC-languages for instance [11, 12, 26–29, 37, 38, 69, 70, 77], have been in use for more than thirty years now. Due to the lack of hardware and instruction set support for interval arithmetic, subroutine calls have to be used by the compiler to map the interval operators and comparisons to appropriate floating-point instructions. This slows down interval arithmetic by a factor close to ten compared to the corresponding floating-point arithmetic.

It has been shown in the last Section that with appropriate hardware support interval operations can be made as fast as floating-point operations. Three additional measures are necessary to let an interval calculation on the computer run at a speed comparable to the corresponding floating-point calculation:

1. Interval arithmetic hardware must be supported by the instruction set of the processor.
2. The high level programming language should provide operators for floating-point operations with directed roundings. The language must provide data types for intervals, and operators for interval operations and comparisons. It must allow overloading of names of elementary functions for interval data types.
3. The compiler must directly map the interval operators, comparisons and elementary functions of the high level programming language onto the instruction set of the processor. This mapping must not be done by slow function or subroutine calls.

From the mathematical point of view the following instructions for interval operations are desirable ($A = [a_1, a_2], B = [b_1, b_2]$):

Algebraic operators:

addition	$C := A + B$	$C := [a_1 \triangledown b_1, a_2 \triangle b_2]$,
subtraction	$C := A - B$	$C := [a_1 \triangledown b_2, a_2 \triangle b_1]$,
negation	$C := -A$	$C := [-a_2, -a_1]$,
multiplication	$C := A * B$	Table 3.6,
division	$C := A/B, 0 \notin B$	Table 3.7,
	$C := A/B, 0 \in B$	Table 3.8,
scalar product	$C := \Diamond(A * B)$	for interval vectors $A = (A_k)$ and $B = (B_k)$, see the first chapter.

Comparisons and lattice operations:

equality	$A = B$	$a_1 = b_1, a_2 = b_2$,
less than or equal	$A \leq B$	$a_1 \leq b_1, a_2 \leq b_2$,
greatest lower bound	$C := glb\,(A, B)$	$C := [\min(a_1, b_1), \min(a_2, b_2)]$,
least upper bound	$C := lub\,(A, B)$	$C := [\max(a_1, b_1), \max(a_2, b_2)]$,
inclusion	$A \subseteq B$	$b_1 \leq a_1, a_2 \leq b_2$,
element of	$a \in A$	$a_1 \leq a \leq a_2$,
interval hull	$C := A \underline{\cup} B$	$C := [\min(a_1, b_1), \max(a_2, b_2)]$,
intersection	$C := A \cap B$	$C :=$ if $\max(a_1, b_1) \leq \min(a_2, b_2)$ then $[\max(a_1, b_1), \min(a_2, b_2)]$ else $[\,]$.

Other comparisons can directly be obtained by comparison of bounds of the intervals A and B. With two comparators all comparisons and lattice operations can be performed in parallel in one time step.

Fast multiple precision arithmetic and fast multiple precision interval arithmetic can easily be obtained by means of the exact scalar product. See Remark 3 on page 60 in section 1.7.

3.10.6 Final Remarks

We have seen that relatively modest hardware equipment consisting of two operation units, a few multiplexers and comparators could make interval arithmetic as fast as floating-point arithmetic. For multiplication various approaches have been compared. The case selection clearly is the most favorable one. Several of the multiplication algorithms use and compare double length products. These are not easily obtainable on most existing processors. Since the double length product is a fundamental operation for other applications as well, like complex arithmetic, the accurate dot product, vector and matrix arithmetic, we require that multipliers should have a fifth rounding mode, namely "unrounded", which enables them to provide the full double length product.

A similar situation appears in interval arithmetic, if division by an interval which contains zero is permitted. The result of an interval division then may consist of two disjoint intervals. In applications of interval arithmetic, division by an interval which contains zero has been used for 30 years now. This forces an extension of the real numbers $I\!R$ by $-\infty$ and $+\infty$ and a consideration of the complete lattice $I\!R^* := I\!R \cup \{-\infty\} \cup \{+\infty\}$ and its computer representable subset $R^* := R \cup \{-\infty\} \cup \{+\infty\}$. In the early days of interval arithmetic attempts were made to define interval arithmetic immediately in $I\!R^*$ instead of $I\!R$. See [23–25] and others. This leads to deep mathematical considerations. We did not follow such lines in this study. The Extended Interval Newton Method is practically the only frequent application where $-\infty$ and $+\infty$ are needed. But there they appear only in an intermediate step as auxiliary values and they disappear immediately in the next step when the intersection with the former approximation is taken.

In conventional numerical analysis Newton's method is the key method for nonlinear problems. The method converges quadratically to the solution if the initial value of the iteration is already close enough. However, it may fail in finite as well as in infinite precision arithmetic even in the case of only a single solution in a given interval. In contrast to this the interval version of Newton's method is globally convergent. It never fails, not even in rounded arithmetic. Newton's method reaches its final elegance and strength in the Extended Interval Newton Method. It encloses all (single) zeros in a given domain. It is locally quadratically convergent. The key operation to achieve these fascinating properties is division by an interval which contains zero. It separates different solutions from each other. A Method which provides for computation of all zeros of a system of nonlinear equations in a given domain is much more frequently applied than the conventional Newton method. This justifies taking division by an interval which contains zero into the basic set of interval operations, and supporting it within the instruction set of the computer.

Bibliography and Related Literature

1. Adams, E.; Kulisch, U.(eds.): **Scientific Computing with Automatic Result Verification.** I. Language and Programming Support for Verified Scientific Computation, II'. Enclosure Methods and Algorithms with Automatic Result Verification, III'. Applications in the Engineering Sciences. Academic Press, San Diego, 1993 (ISBN 0-12-044210-8).
2. Albrecht, R.; Alefeld, G.; Stetter, H.J. (Eds.): **Validation Numerics – Theory and Applications.** Computing Supplementum **9**, Springer-Verlag, Wien / New York, 1993.
3. Alefeld, G.: *Intervallrechnung über den komplexen Zahlen und einige Anwendungen.* Dissertation, Universität Karlsruhe, 1968.
4. Alefeld, G.: *Über die aus monoton zerlegbaren Operatoren gebildeten Iterationsverfahren.* Computing **6**, pp. 161-172, 1970.
5. Alefeld, G.; Herzberger, J.: **Einführung in die Intervallrechnung.** Bibliographisches Institut (Reihe Informatik, Nr. 12), Mannheim / Wien / Zürich, 1974 (ISBN 3-411-01466-0).
6. Alefeld, G.; Herzberger, J.: **An Introduction to Interval Computations.** Academic Press, New York, 1983 (ISBN 0-12-049820-0).
7. Alefeld, G.; Mayer, G.: *Einschließungsverfahren.* In [22, pp. 155-186], 1995.
8. Alefeld, G.; Frommer, A.; Lang, B. (eds.): **Scientific Computing and Validated Numerics.** Proceedings of SCAN-95. Akademie Verlag, Berlin, 1996. ISBN 3-05-501737-4
9. Baumhof, Ch.: *Ein Vektorarithmetik-Koprozessor in VLSI-Technik zur Unterstützung des Wissenschaftlichen Rechnens.* Dissertation, Universität Karlsruhe, 1996.
10. Blomquist, F.: **PASCAL-XSC, BCD-Version 1.0, Benutzerhandbuch für das dezimale Laufzeitsystem.** Institut für Angewandte Mathematik, Universität Karlsruhe, 1997.
11. Bohlender, G.; Rall, L. B.; Ullrich, Ch.; Wolff v. Gudenberg, J.: **PASCAL–SC: Wirkungsvoll programmieren, kontrolliert rechnen.** Bibliographisches Institut, Mannheim / Wien / Zürich, 1986 (ISBN 3-411-03113-1).
12. Bohlender, G.; Rall, L. B.; Ullrich, Ch.; Wolff v. Gudenberg, J.: **PASCAL–SC: A Computer Language for Scientific Computation.** Perspectives in Computing, Vol. 17, Academic Press, Orlando, 1987 (ISBN 0-12-111155-5).
13. Bohlender, G.: *Literature on Enclosure Methods and Related Topics.* Institut für Angewandte Mathematik, Universität Karlsruhe, pp. 1-68, 2000.
14. Collatz, L.: **Funktionalanalysis und numerische Mathematik.** Springer–Verlag, Berlin / Heidelberg / New York, 1968.
15. Fischer, H.-C.: *Schnelle automatische Differentiation, Einschließungsmethoden und Anwendungen.* Dissertation, Universität Karlsruhe, 1990.
16. Fischer, H.: *Automatisches Differenzieren.* In [22, pp. 53-104], 1995.

17. Hammer, R.; Hocks, M.; Kulisch, U.; Ratz, D.: **Numerical Toolbox for Verified Computing I: Basic Numerical Problems.** (Vol. II see [31], version in C++ see [18]) Springer–Verlag, Berlin / Heidelberg / New York, 1993.
18. Hammer, R.; Hocks, M.; Kulisch, U.; Ratz, D.: **C++ Toolbox for Verified Computing: Basic Numerical Problems.** Springer–Verlag, Berlin / Heidelberg / New York, 1995.
19. Hansen, E.: **Topics in Interval Analysis.** Clarendon Press, Oxford, 1969.
20. Hansen, E.: **Global Optimization Using Interval Analysis.** Marcel Dekker Inc., New York/Basel/Hong Kong, 1992.
21. Herzberger, J. (ed.): **Topics in Validated Computations.** Proceedings of IMACS–GAMM International Workshop on Validated Numerics, Oldenburg, 1993. North Holland, 1994.
22. Herzberger, J.: **Wissenschaftliches Rechnen, Eine Einführung in das Scientific Computing.** Akademie Verlag, 1995.
23. Kaucher, E.: *Über metrische und algebraische Eigenschaften einiger beim numerischen Rechnen auftretender Räume.* Dissertation, Universität Karlsruhe, 1973.
24. Kaucher, E.: *Algebraische Erweiterungen der Intervallrechnung unter Erhaltung der Ordnungs- und Verbandsstrukturen.* In: Albrecht, R.; Kulisch, U. (Eds.): **Grundlagen der Computerarithmetik.** Computing Supplementum **1.** Springer-Verlag, Wien / New York, pp. 65-79, 1977.
25. Kaucher, E.: *Über Eigenschaften und Anwendungsmöglichkeiten der erweiterten Intervallrechnung und des hyperbolischen Fastkörpers über* **R**. In: Albrecht, R.; Kulisch, U. (Eds.): **Grundlagen der Computerarithmetik.** Computing Supplementum **1.** Springer-Verlag, Wien / New York, pp. 81-94, 1977.
26. Klatte, R.; Kulisch, U.; Neaga, M.; Ratz, D.; Ullrich, Ch.: **PASCAL–XSC — Sprachbeschreibung mit Beispielen.** Springer-Verlag, Berlin/Heidelberg/New York, 1991 (ISBN 3-540-53714-7, 0-387-53714-7).
27. Klatte, R.; Kulisch, U.; Neaga, M.; Ratz, D.; Ullrich, Ch.: **PASCAL–XSC — Language Reference with Examples.** Springer-Verlag, Berlin/Heidelberg/New York, 1992.
28. Klatte, R.; Kulisch, U.; Lawo, C.; Rauch, M.; Wiethoff, A.: **C–XSC, A C++ Class Library for Extended Scientific Computing.** Springer-Verlag, Berlin/Heidelberg/New York, 1993.
29. Klatte, R.; Kulisch, U.; Neaga, M.; Ratz, D.; Ullrich, Ch.: **PASCAL–XSC — Language Reference with Examples (In Russian).** Moscow, 1994, second edition 2000.
30. Knöfel, A.: *Hardwareentwurf eines Rechenwerks für semimorphe Skalar- und Vektoroperationen unter Berücksichtigung der Anforderungen verifizierender Algorithmen.* Dissertation, Universität Karlsruhe, 1991.
31. Krämer, W.; Kulisch, U.; Lohner, R.: **Numerical Toolbox for Verified Computing II: Theory, Algorithms and Pascal-XSC Programs.** (Vol. I see [17, 18]) Springer–Verlag, Berlin / Heidelberg / New York, to appear.
32. Krawczyk, R.; Neumaier, A.: *Interval Slopes for Rational Functions and Associated Centered Forms.* SIAM Journal on Numerical Analysis **22**, pp. 604-616, 1985.
33. Kulisch, U.: **Grundlagen des Numerischen Rechnens — Mathematische Begründung der Rechnerarithmetik.** Reihe Informatik, Band 19, Bibliographisches Institut, Mannheim/Wien/Zürich, 1976 (ISBN 3-411-01517-9).
34. Kulisch, U.; Miranker, W. L.: **Computer Arithmetic in Theory and Practice.** Academic Press, New York, 1981 (ISBN 0-12-428650-x).

35. Kulisch, U.; Ullrich, Ch. (Eds.): **Wissenschaftliches Rechnen und Programmiersprachen.** Proceedings of Seminar held in Karlsruhe, April 2–3, 1982. Berichte des German Chapter of the ACM, Band 10, B. G. Teubner Verlag, Stuttgart, 1982 (ISBN 3-519-02429-2).
36. Kulisch, U.; Miranker, W. L. (Eds.): **A New Approach to Scientific Computation.** Proceedings of Symposium held at IBM Research Center, Yorktown Heights, N. Y., 1982. Academic Press, New York, 1983 (ISBN 0-12-428660-7).
37. Kulisch, U. (Ed.): **PASCAL–SC: A PASCAL extension for scientific computation**, Information Manual and Floppy Disks, Version IBM PC/AT; Operating System DOS'. B. G. Teubner Verlag (Wiley-Teubner series in computer science), Stuttgart, 1987 (ISBN 3-519-02106-4 / 0-471-91514-9).
38. Kulisch, U. (Ed.): **PASCAL–SC: A PASCAL extension for scientific computation**, Information Manual and Floppy Disks, Version ATARI ST'. B. G. Teubner Verlag, Stuttgart, 1987 (ISBN 3-519-02108-0).
39. Kulisch, U. (Ed.): **Wissenschaftliches Rechnen mit Ergebnisverifikation — Eine Einführung.** Ausgearbeitet von S. Geörg, R. Hammer und D. Ratz. Vol. 58. Akademie Verlag, Berlin, und Vieweg Verlagsgesellschaft, Wiesbaden, 1989.
40. Kulisch, U.: *Advanced Arithmetic for the Digital Computer — Design of Arithmetic Units.* Electronic Notes in Theoretical Computer Science, http://www.elsevier.nl/locate/entcs/volume24.html pp. 1-72, 1999.
41. Lohner, R.: *Einschließung der Lösung gewöhnlicher Anfangs- und Randwertaufgaben und Anwendungen.* Dissertation, Universität Karlsruhe, 1988.
42. Lohner, R.: *Computation of Guaranteed Enclosures for the Solutions of Ordinary Initial and Boundary Value Problems.* pp. 425–435 in: Cash, J. R.; Gladwell, I. (Eds.): **Computational Ordinary Differential Equations.** Clarendon Press, Oxford, 1992.
43. Mayer, G.: *Grundbegriffe der Intervallrechnung.* In [39, pp. 101-117], 1989.
44. Moore, R. E.: **Interval Analysis.** Prentice Hall Inc., Englewood Cliffs, N. J.; 1966.
45. Moore, R. E.: **Methods and Applications of Interval Analysis.** SIAM, Philadelphia, Pennsylvania, 1979.
46. Moore, R. E. (Ed.): **Reliability in Computing: The Role of Interval Methods in Scientific Computing.** Proceedings of the Conference at Columbus, Ohio, September 8–11, 1987; Perspectives in Computing **19**, Academic Press, San Diego, 1988 (ISBN 0-12-505630-3).
47. Neumaier, A.: **Interval Methods for Systems of Equations.** Cambridge University Press, Cambridge, 1990.
48. Neumann, J. von; Goldstine, H. H.: *Numerical Inverting of Matrices of High Order.* Bulletin of the American Mathematical Society, 53, 11, pp. 1021-1099, 1947.
49. Rall, L. B.: **Automatic Differentiation: Techniques and Applications.** Lecture Notes in Computer Science, No. 120, Springer-Verlag, Berlin, 1981.
50. Ratschek, H.; Rokne, J.: **Computer Methods for the Range of Functions.** Ellis Horwood Limited, Chichester, 1984.
51. Ratz, D.: *Programmierpraktikum mit PASCAL–SC.* In: Höhler, G.; Staudenmaier, H. M. (Hrsg.): **Computer Theoretikum und Praktikum für Physiker.** Band **5**, Fachinformationszentrum Karlsruhe, 1990.
52. Ratz, D.: *Globale Optimierung mit automatischer Ergebnisverifikation.* Dissertation, Universität Karlsruhe, 1992.
53. Ratz, D.: **Automatic Slope Computation and its Application in Nonsmooth Global Optimization.** Shaker Verlag, Aachen, 1998.

54. Ratz, D.: *On Extended Interval Arithmetic and Inclusion Isotony.* Preprint, Institut für Angewandte Mathematik, Universität Karlsruhe, 1999.
55. Rump, S. M.: *Kleine Fehlerschranken bei Matrixproblemen.* Dissertation, Universität Karlsruhe, 1980.
56. Rump, S. M.: *How Reliable are Results of Computers? / Wie zuverlässig sind die Ergebnisse unserer Rechenanlagen?* In: *Jahrbuch Überblicke Mathematik*, Bibliographisches Institut, Mannheim, 1983.
57. Rump, S.M.: *Validated Solution of Large Linear Systems.* In [2, pp. 191-212], 1993.
58. Rump, S.M.: *Verification Methods for Dense and Sparse Systems of Equations.* In [21, pp. 63-135], 1994.
59. Rump, S.M.: *INTLAB – Interval Laboratory.* TU Hamburg-Harburg, 1998.
60. Schmidt, L.: *Semimorphe Arithmetik zur automatischen Ergebnisverifikation auf Vektorrechnern.* Dissertation, Universität Karlsruhe, 1992.
61. Shiriaev, D. V.: *Fast Automatic Differentiation for Vector Processors and Reduction of the Spatial Complexity in a Source Translation Environment.* Dissertation, Universität Karlsruhe, 1994.
62. Sunaga, T.: *Theory of an interval algebra and its application to numerical analysis.* RAAG Memoires 2, pp. 547-564, 1958.
63. Teufel, T.: *Ein optimaler Gleitkommaprozessor.* Dissertation, Universität Karlsruhe, 1984.
64. Ullrich, Ch. (Ed.): **Computer Arithmetic and Self-Validating Numerical Methods.** (Proceedings of SCAN 89, held in Basel, Oct. 2-6, 1989, invited papers). Academic Press, San Diego, 1990.
65. Walter, W. V.: *FORTRAN–SC, A FORTRAN Extension for Engineering / Scientific Computation with Access to ACRITH: Language Description with Examples.* In [46, pp. 43-62], 1988.
66. Walter, W. V.: *Einführung in die wissenschaftlich-technische Programmiersprache FORTRAN–SC.* ZAMM **69**, 4, T52-T54, 1989.
67. Walter, W. V.: *FORTRAN–SC: A FORTRAN Extension for Engineering / Scientific Computation with Access to ACRITH, Language Reference and User's Guide.* 2nd ed., pp. 1-396, IBM Deutschland GmbH, Stuttgart, Jan. 1989.
68. Walter, W. V.: *Flexible Precision Control and Dynamic Data Structures for Programming Mathematical and Numerical Algorithms.* Dissertation, Universität Karlsruhe, 1990.
69. Wippermann, H.-W.: *Realisierung einer Intervallarithmetik in einem ALGOL-60 System.* Elektronische Rechenanlagen **9**, pp. 224-233, 1967.
70. Wippermann, H.-W.: *Implementierung eines ALGOL-60 Systems mit Schrankenzahlen.* Elektronische Datenverarbeitung **10**, pp. 189-194, 1968.
71. Wolff v. Gudenberg, J.: *Hardware Support for Interval Arithmetic, Extended Version.* Report No. 125, Institut für Informatik, Universität Würzburg, 1995.
72. Wolff v. Gudenberg, J.: *Hardware Support for Interval Arithmetic.* In [8, pp. 32-38], 1996.
73. Wolff v. Gudenberg, J.: **Proceedings of Interval'96.** International Conference on Interval Methods and Computer Aided Proofs in Science and Engineering, Würzburg, Germany, Sep. 30 - Oct. 2, 1996. Special issue 3/97 of the journal Reliable Computing, 1997.
74. Yohe, J.M.: *Roundings in Floating-Point Arithmetic.* IEEE Trans. on Computers, Vol. C-22, No. 6, June 1973, pp. 577-586.
75. IBM: *IBM System/370 RPQ'. High Accuracy Arithmetic.* SA 22-7093-0, IBM Deutschland GmbH (Department 3282, Schönaicher Strasse 220, D-71032 Böblingen), 1984.

Bibliography and Related Literature   141

76. IBM: **IBM High-Accuracy Arithmetic Subroutine Library (ACRITH).** IBM Deutschland GmbH (Department 3282, Schönaicher Strasse 220, D-71032 Böblingen), 3rd edition, 1986.
   1. General Information Manual. GC 33-6163-02.
   2. Program Description and User's Guide. SC 33-6164-02.
   3. Reference Summary. GX 33-9009-02.
77. IBM: **ACRITH–XSC: IBM High Accuracy Arithmetic — Extended Scientific Computation. Version 1, Release 1.** IBM Deutschland GmbH (Schönaicher Strasse 220, D-71032 Böblingen), 1990.
   1. General Information, GC33-6461-01.
   2. Reference, SC33-6462-00.
   3. Sample Programs, SC33-6463-00.
   4. How To Use, SC33-6464-00.
   5. Syntax Diagrams, SC33-6466-00.
78. American National Standards Institute / Institute of Electrical and Electronics Engineers: *A Standard for Binary Floating-Point Arithmetic.* ANSI/IEEE Std. 754-1985, New York, 1985 (reprinted in SIGPLAN **22**, 2, pp. 9-25, 1987). Also adopted as IEC Standard 559:1989.
79. American National Standards Institute / Institute of Electrical and Electronics Engineers: *A Standard for Radix-Independent Floating-Point Arithmetic.* ANSI/IEEE Std. 854-1987, New York, 1987.
80. IMACS; GAMM: *IMACS-GAMM Resolution on Computer Arithmetic.* In Mathematics and Computers in Simulation **31**, pp. 297-298, 1989. In Zeitschrift für Angewandte Mathematik und Mechanik **70**, no. 4, p. T5, 1990.
81. IMACS; GAMM: *GAMM-IMACS Proposal for Accurate Floating-Point Vector Arithmetic.* GAMM, Rundbrief 2, pp. 9-16, 1993. Mathematics and Computers in Simulation, Vol. **35**, IMACS, North Holland, 1993. News of IMACS, Vol. 35, No. 4, pp. 375-382, Oct. 1993.
82. SIEMENS: **ARITHMOS (BS 2000) Unterprogrammbibliothek für Hochpräzisionsarithmetik. Kurzbeschreibung, Tabellenheft, Benutzerhandbuch.** SIEMENS AG, Bereich Datentechnik, Postfach 83 09 51, D-8000 München 83. Bestellnummer U2900-J-Z87-1, Sept. 1986.
83. Sun Microsystems: **Interval Arithmetic Programming Reference, Fortran 95.** Sun Microsystems, Inc., 901 San Antonio Road, Palo Alto, CA 94303, USA, 2000.

SpringerMathematics

Ulrich Kulisch,
Rudolf Lohner,
Axel Facius (eds.)

Perspectives on Enclosure Methods

2001. XII, 345 pages. With numerous figures.
Softcover EUR 59,90
(Recommended retail price)
All prices are net-prices subject to local VAT,
Net-price subject to local VAT.
ISBN 3-211-83590-3

Enclosure methods and their applications have been developed to a high standard during the last decades. These methods guarantee the validity of the computed results, this means they are of the same standard as the rest of mathematics. This book deals with a wide variety of aspects of enclosure methods.

All contributions follow the common goal to push the limits of enclosure methods forward. Topics that are treated include basic questions of arithmetic, proving conjectures, bounds for Krylow type linear system solvers, bounds for eigenvalues, the wrapping effect, algorithmic differencing, differential equations, finite element methods, application in robotics, and nonsmooth global optimization.

A-1201 Wien, Sachsenplatz 4–6, P.O. Box 89, Fax +43.1.330 24 26, e-mail: books@springer.at, Internet: www.springer.at
D-69126 Heidelberg, Haberstraße 7, Fax +49.6221.345-229, e-mail: orders@springer.de
USA, Secaucus, NJ 07096-2485, P.O. Box 2485, Fax +1.201.348-4505, e-mail: orders@springer-ny.com
Eastern Book Service, Japan, Tokyo 113, 3–13, Hongo 3-chome, Bunkyo-ku, Fax +81.3.38 18 08 64, e-mail: orders@svt-ebs.co.jp

Springer-Verlag
and the Environment

WE AT SPRINGER-VERLAG FIRMLY BELIEVE THAT AN international science publisher has a special obligation to the environment, and our corporate policies consistently reflect this conviction.

WE ALSO EXPECT OUR BUSINESS PARTNERS – PRINTERS, paper mills, packaging manufacturers, etc. – to commit themselves to using environmentally friendly materials and production processes.

THE PAPER IN THIS BOOK IS MADE FROM NO-CHLORINE pulp and is acid free, in conformance with international standards for paper permanency.